Springer Theses

Recognizing Outstanding Ph.D. Research

For further volumes:
http://www.springer.com/series/8790

Aims and Scope

The series "Springer Theses" brings together a selection of the very best Ph.D. theses from around the world and across the physical sciences. Nominated and endorsed by two recognized specialists, each published volume has been selected for its scientific excellence and the high impact of its contents for the pertinent field of research. For greater accessibility to non-specialists, the published versions include an extended introduction, as well as a foreword by the student's supervisor explaining the special relevance of the work for the field. As a whole, the series will provide a valuable resource both for newcomers to the research fields described, and for other scientists seeking detailed background information on special questions. Finally, it provides an accredited documentation of the valuable contributions made by today's younger generation of scientists.

Theses are accepted into the series by invited nomination only and must fulfill all of the following criteria

- They must be written in good English.
- The topic should fall within the confines of Chemistry, Physics, Earth Sciences, Engineering and related interdisciplinary fields such as Materials, Nanoscience, Chemical Engineering, Complex Systems and Biophysics.
- The work reported in the thesis must represent a significant scientific advance.
- If the thesis includes previously published material, permission to reproduce this must be gained from the respective copyright holder.
- They must have been examined and passed during the 12 months prior to nomination.
- Each thesis should include a foreword by the supervisor outlining the significance of its content.
- The theses should have a clearly defined structure including an introduction accessible to scientists not expert in that particular field.

Anna Alexandra Vackiner

Sedimentary Facies Reconstruction and Kinematic Restoration of Tight Gas Fields

Studies from the Upper Permian in Northwestern Germany

Doctoral Thesis accepted by
the RWTH Aachen University, Germany

Springer

Author
Dr. Anna Alexandra Vackiner
Energy and Mineral Resources Group
Geological Institute
RWTH Aachen University
Aachen
Germany

Supervisors
Prof. Dr. Peter A. Kukla
Energy and Mineral Resources Group
Geological Institute
RWTH Aachen University
Aachen
Germany

Prof. Dr. Harald Stollhofen
North Bavarian Center of Earth Sciences
Friedrich-Alexander University
Erlangen–Nürnberg
Germany

ISSN 2190-5053 ISSN 2190-5061 (electronic)
ISBN 978-3-642-36045-9 ISBN 978-3-642-36047-3 (eBook)
DOI 10.1007/978-3-642-36047-3
Springer Heidelberg New York Dordrecht London

Library of Congress Control Number: 2013931748

Printed on acid-free paper

Springer is part of Springer Science+Business Media (www.springer.com)

Parts of this thesis have been published in the following journal articles:

Anna Alexandra Vackiner, Philipp Antrett, Harald Stollhofen, Stefan Back, Peter Alfred Kukla and Claudia Bärle (2011) Syndepositional Tectonic Controls and Palaeo-Topography of a Permian Tight Gas Reservoir in NW Germany. Journal of Petroleum Geology, Vol. 34(4), p. 411–428.

Anna Alexandra Vackiner, Philipp Antrett, Frank Strozyk, Harald Stollhofen, Stefan Back and Peter Alfred Kukla (2012) Reconstructing the Upper Permian sedimentary facies distribution of a tight gas field in Central Europe on the basis of a modern analog field study in the Panamint Valley, western U.S., Geosphere, Vol. 8(5), p. 1129–1145.

Anna Alexandra Vackiner, Philipp Antrett, Frank Strozyk, Harald Stollhofen, Peter Kukla, Stefan Back (accepted) Integrating salt kinematics and regional tectonics across a Permian gas field: a case study from East Frisia, NW Germany. International Journal of Earth Sciences.

When you reach the end of your rope, tie a knot in it and hang on

—Thomas Jefferson

Supervisor's Foreword

Natural Hydrocarbon Gas is considered to represent the "bridging fuel" until new energies become technically and economically viable. Among hydrocarbon gas, one can classify conventional, unconventional and tight gas resources. The unconventional and tight gas resources have recently received much interest because of the potential very large reserves, which could be produced with suitable technology. Tight gas reservoirs which are found throughout the world and which occur in all common types of reservoir have been produced for many decades, but still pose a major technical challenge owing to their heterogeneous reservoir characteristics and in particular their low permeability and low porosity. Given the global importance of such reservoirs, the understanding of the complexity of tight gas fields therefore requires an integrated approach involving geological, geophysical and petrophysical analysis which very few publications to date have achieved.

The thesis of Alexandra presents such a multidisciplinary approach towards the analysis of complex tight gas reservoirs using the example of the Permian in the Northwest European Basin. Despite a long exploration and production history in this basin, previous work in Germany has mainly concentrated on the overall basin evolution based on seismic and wireline borehole data and the diagenesis of the reservoirs based on core data. Very little work on the other hand has been undertaken to discern the multiple reactivations of structural elements within the basin and to separate syntectonic versus post-tectonic deformation patterns and styles. The methods used in this work therefore include geological and geophysical analyses, remote sensing, 2D retro-deformation modelling and log- and core analysis.

In a first step, a detailed and high-resolution facies analysis of fluvio-eolian sediments of the Upper Rotliegend II sequences was undertaken. A separation of different dune types and interdune deposits was achieved and supported by a comparison with modern reservoirs in Panamint Valley, western USA leading to a palaeo-sedimentary and palaeo-geomorphologic model through time. A sequential retro-deformation of the study area aimed at a differentiation between salt rise mechanisms, their timing and their relation to regional tectonic events. It involves sedimentation, decompaction, fault-related deformation, salt movement, thermal subsidence and isostasy. Results suggest that reactive diapirism started with lateral salt movement and local injection of salt into the overburden during the Lower Triassic Buntsandstein deposition. During the Upper Buntsandstein, rafts developed and small diapirs first breached through the sediment surface. Active diapirism during the Upper Triassic (Lower Keuper) was accompanied by salt piercement of the overburden. For the Jurassic, hydrothermal fluid circulation along active faults in an extensional regime is postulated. In the Lower Cretaceous, the salt rise mechanism changed to passive diapirism, which is still continuous until today. The downbuilding phase was accompanied by the development of large salt rim synclines leading to local development of faults. The structural analysis further shows multiple reactivations which demonstrate that syn- and post-sedimentary tectonics exert a strong control on facies' architecture in the basins. The importance of fault tip geometries and pull-apart structures for sedimentation and fluid flow could further be established.

Alexandras's research formed part of the Tight Gas Initiative (TGI) between RWTH Aachen University and Wintershall Holding GmbH which supported this study. She has presented her work in several international publications and at international conferences. Her thesis is a well-documented piece of work which presents for the first time an integrated model in which the influence of regional tectonics, salt tectonics, sedimentation, compaction, thermal subsidence and isostasy on tight gas reservoir evolution and distribution can be discerned. It therefore represents a major step forward in the understanding of this complex reservoir type.

Aachen, September 2012 Prof. Dr. Peter A. Kukla

Acknowledgments

My sincerest thanks go to my advisors Prof. Peter Kukla Ph.D. and Prof. Dr. Harald Stollhofen. I thank Peter Kukla for his supervision and for protecting and defending me when scientific disagreements occurred. I thank Prof. Dr. Harald Stollhofen for sharing his enormous knowledge about sedimentology and for the time he invested for detailed internal reviews of my manuscripts. I am grateful to my colleagues, co-authors and friends Philipp Antrett, Dr. Frank Strozyk, Dr. Stefan Back and Dr. Chris Hilgers for long discussions, logistic support and high spirits. Special thanks to Frank, who invested a lot of time in cross-reading my manuscripts. To my additional co-authors, Vanessa Havenith, Dr. Claudia Bärle, Dr. Sven Sindern, Prof. Dr. Michael Meyer and Dr. Ina Blumenstein-Weingartz, I am deeply grateful.

This doctoral thesis is part of the Wintershall and RWTH Aachen University Tight-Gas Initiative. I thank the Wintershall Holding GmbH and the GDF Suez E&P Deutschland GmbH for providing the seismic data and the core material and for financing the project. The thesis benefitted from fruitful discussions with the industrial partners during several meetings. In particular Dr. Claudia Bärle, Dr. Harald Karg, Bernhard Siethoff, Dr. Wolfram Unverhaun, Petra Unverhaun, Michael Blum, Dr. Wolf-Dieter Karnin, Dr. Ina Blumenstein-Weingartz, Dr. Dirk Adelmann, Jan Himmerkus and Dr. Dieter Kaufmann (all from Wintershall Holding GmbH), Dr. Robert Bussert and Dr. Anton Irmen (GDF Suez E&P Deutschland GmbH) shared their knowledge and experience.

Concerning the fieldwork, I thank Dr. Norbert Klitzsch of the Applied Geophysics and Geothermal Energy Department of the E.ON Energy Research Center, who gave Philipp Antrett and me the opportunity to conduct ground resistivity measurements in the Panamint Valley, and Rebecca Möller, who took care of the bedevilled shipping of the ground resistivity equipment from Germany to the US and back. Without Rebecca, the US or the German customs could still call a complete ground-resistivity equipment their own. Furthermore, I would like to thank the United States National Park Services, especially Richard Friese, for issuing a Research and Collecting Permit for the Death Valley National Park. He also gave me the opportunity to ship a potential dinosaur egg to Germany, which, unfortunately, turned out to be a huge pebble. I thank Dr. Marco Möller for taking care of our suitable lodging during fieldwork with lending one of the Hilleberg tents of the glacial research group of the Geographical Department. It stood strong desert winds and a devastating sandstorm, during which several other tents collapsed.

Further, I thank the volume editor of the Journal of Petroleum Geology Christopher Tiratsoo and the reviewer Prof. Dr. Nigel Mountney for their constructive support, which greatly improved the first of my scientific papers (largely embedded in this thesis; Chap. 4).

Finally, I would like to thank my friends and family, especially Susan Giffin and Beke Rosleff-Sörensen, for being the best circuit training and athletics companions I ever had, Philipp Antrett for his geological and mental support and for his love, and my parents, Gerhard and Christa Vackiner, for the emotional support, their love and care.

Contents

1 Introduction 1
1.1 Rationale 1
1.2 Objectives. 2
1.3 Thesis Outline. 4
References 5

2 Geological Setting 7
2.1 Regional Geological Setting 7
2.2 Geology of the Study Area. 10
References 11

3 Data and Methods 13
3.1 Data and Methods 13
3.2 Methods: Sedimentary Facies Analysis from Core Material 16
3.2.1 Pond/Lake 17
3.2.2 Pond/Lake Margin. 18
3.2.3 Aeolian Mudflat 19
3.2.4 Wet to Damp Sandflat. 21
3.2.5 Dry Sandflat. 22
3.2.6 Aeolian Dune (base) 23
3.2.7 Low Energetic Meandering Fluvial Deposits 24
3.2.8 High Energetic Braided Stream Fluvial Deposits 26
References 29

4 Syndepositional Tectonic Controls and Palaeo-Topography of a Permian Tight Gas Reservoir in NW Germany 31
4.1 Introduction 31
4.1.1 Geological Framework. 32
4.2 Data and Interpretation Methodology 34
4.3 Structural Subdivision of the Study Area 35
4.4 Palaeo-Relief Analysis from Seismic Data 38
4.5 Palaeo-Relief Analysis from Core and Log Data 42
4.6 Palaeo-Relief Uncertainty During Zechstein 43

4.7 Discussion 45
4.8 Conclusions 49
References 51

5 **The Panamint Valley, Western US: A Field Analogue for the Sedimentary Facies Distribution of a Permian Tight Gas Field in Central Europe** 55
5.1 Introduction 55
5.2 Setting 58
5.2.1 Geological Setting of Panamint Valley 58
5.2.2 Subsurface Area Germany 59
5.3 Data and Methods 60
5.4 Results 61
5.4.1 Panamint Valley 61
5.4.2 Subsurface Study Area Germany 66
5.5 Discussion 69
5.6 Conclusions 71
References 72

6 **Integrating Salt Kinematics and Diagenesis in a Tight Gas Field: A Case Study from the Upper Rotliegend in East Frisia** 75
6.1 Introduction 75
6.2 Data and Methods 76
6.2.1 3D Isopach Data 79
6.2.2 2D Retro-Deformation 81
6.3 Results 86
6.3.1 3D Isopach Analysis and Lithologies 87
6.3.2 2D Retro-Deformation 90
6.3.3 Core Data (Sandstone Petrography, Fluid Inclusions) . . . 91
6.4 Interpretation 92
6.5 Conclusions 97
References 98

7 **Facies Analysis from Well Cores, Northern Central Germany: Comparison to NW German Well Cores** 103
7.1 Introduction 103
7.2 Core Analysis Results 108
7.2.1 Well I 108
7.2.2 Well II 108
7.2.3 Well III 109
7.2.4 Well IV 111
7.3 Interpretation 111
7.4 Comparison to North–Western German Study Area 113
References 115

8 Conclusions, Synopsis and Perspectives 117
8.1 Conclusions 117
8.2 Synopsis 121
8.3 Perspectives 121
References 122

Chapter 1
Introduction

1.1 Rationale

The economic development of tight gas reserves plays an increasing role in the hydrocarbon exploration and production in Germany. Large tight gas reserves are produced with suitable technologies such as horizontal drilling and hydraulic fracturing. The reasons for low permeabilities in sandstone reservoirs are manifold and controlled by several parameters and processes. Besides sedimentary characteristics and burial depth, the tectonic and diagenetic development of the gas reservoir sandstones plays a crucial role. The estimation of recoverable tight gas reserves in the North German Basin ranges 100–150 billion cubic metres (BCM) [1]. The term *tight gas reservoir* is not uniformly defined. Law and Curtis [2] defined low permeability sandstone reservoirs as those exhibiting permeabilities <0.1 mD. This value has been committed since as a political margin in the United States to determine which wells deserve governmental funding for gas production [3]. In contrast, the German Society for Petroleum and Coal Science and Technology (DGMK) refers to a tight gas sandstone reservoir if permeabilities are <0.6 mD (Fig. 1.1). After Holditch [3] tight gas reservoirs are gas formations that will not flow on their own after perforation. For economic production a stimulation technique is required.

This doctoral thesis is part of the Tight Gas Initiative (TGI) of the RWTH Aachen University and the Wintershall Holding GmbH, which focuses on multiple parameters of tight gas reservoir development and quality. Previous studies on tight gas research have mainly concentrated on individual aspects. Until now, the integration of spatial and time evolution has not been considered. The overall project motivation is to establish innovative and vigorous concepts for reliable reservoir quality prediction.

A. A. Vackiner, *Sedimentary Facies Reconstruction and Kinematic Restoration of Tight Gas Fields*, Springer Theses, DOI: 10.1007/978-3-642-36047-3_1,

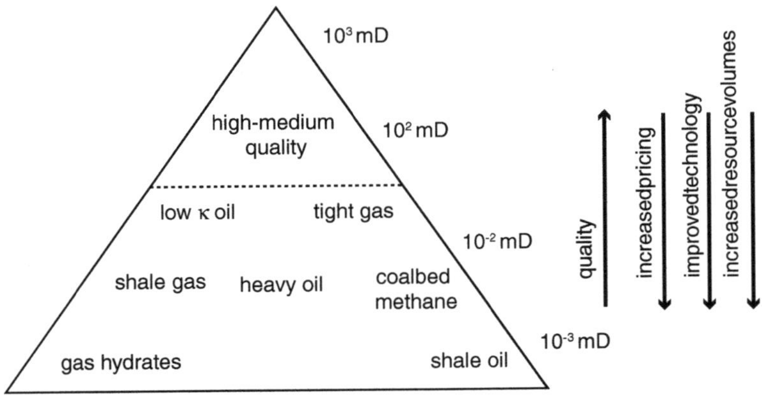

Fig. 1.1 Resource triangle. Modified from Holditch [3, 4]

1.2 Objectives

This study focuses on the development of an Upper Rotliegend II tight gas field through time. The tight gas field is located in East Frisia, north-western Germany. The reservoir rock deposition took place at the south-western margin of the Southern Permian Basin (SPB), which occupied large parts of northern continental Europe and the southern North Sea during the Upper Permian. Reservoir rocks are of fluvio-aeolian origin (Fig. 1.2). The key aim of this doctoral thesis is to provide a documentation of fault zone activity through time and to unravel the relationship between structural geometry, sedimentology and diagenesis. The analysis of the fault zone activity through time provides crucial information about syntectonic sedimentary facies distribution, especially during the Upper Rotliegend II, and about the timing of fluid circulation. One major aim was to develop an Upper Rotliegend II sedimentological model of the reservoir sandstone‘s palaeo-depositional setting with respect to fault-induced palaeo-relief. For a better understanding of the interaction between sedimentology and tectonics a field analogue study in the Panamint Valley, California, United States, has been carried out.

A second study area is located in northern central Germany. The study focuses on the deposits overlying the main reservoir intervals. In terms of the development of unconventional reservoirs the gas bearing heterogeneous clay-dominated lithologies arouse interest.

The following questions, which occurred during the 3 year study, form the framework of this thesis:

How was the Upper Rotliegend II fault-induced palaeo-topography arranged prior to later multi-phase tectonic overprint?

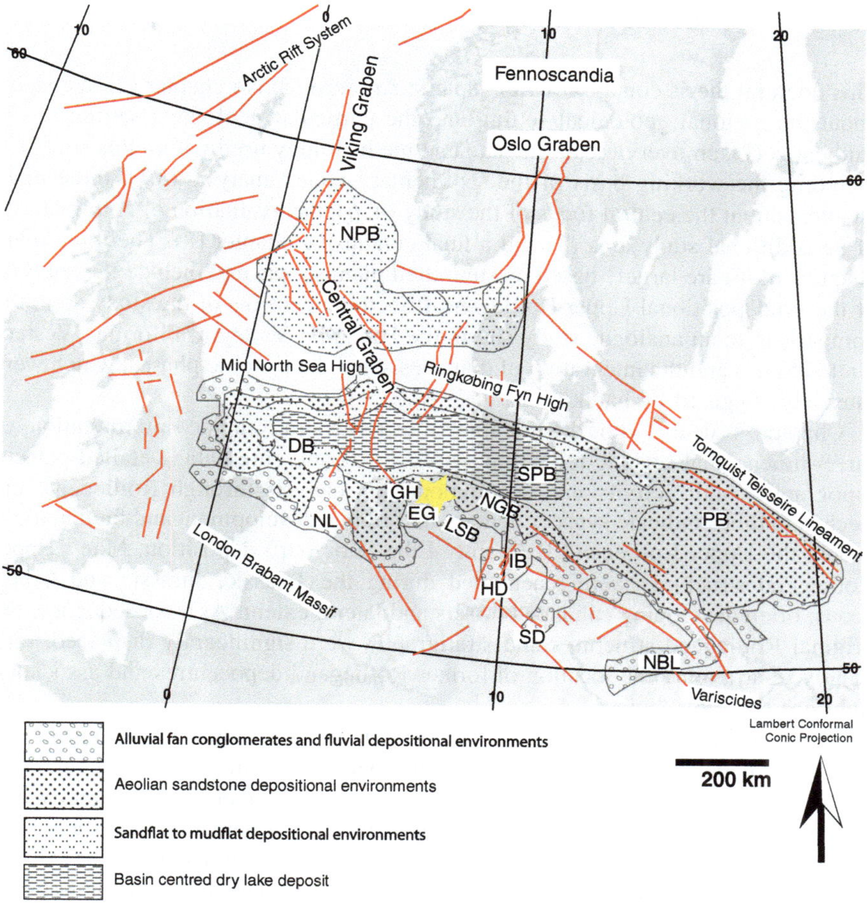

Fig. 1.2 Map outlining maximum extent of depositional areas of the Southern Permian Basin (SPB) and Northern Permian Basin (NPB) during the late Upper Rotliegend II (modified from [5, 6]). *DB* = Dutch Basin, *NL* = Netherlands Low, *GH* = Groningen High, *EG* = Ems Graben, *LSB* = Lower Saxony Basin, *NGB* = North German Basin, *HD* = Hessian Depression, *IB* = Ilfeld Basin, *SD* = Saale Depression, *NBL* = Neiße-Bobr Low, *PB* = Polish Basin, *SPB* = Southern Permian Basin, *NPB* = Northern Permian Basin

Where are the synsedimentary Upper Rotliegend II structural traps for the accumulation of reservoir rocks located in the study area?
Can the Upper Rotliegend II sedimentary facies distribution be reconstructed using a field analogue study?
What role does halokinesis and multi-phase tectonic overprinting play during reservoir rock development?
Are the developed methods applicable to geologically similar study areas?

1.3 Thesis Outline

This doctoral thesis contains eight chapters: An introductive chapter (1), a chapter about the regional geological setting and the particular geological setting of the study area (2), an overview of the data and methodology used within this study (3) including the working basis of the sedimentary facies analysis (3.1), three main chapters about the central focus of the study (4–6), the evaluation of core material of an additional study area (7) and a final conclusive chapter (8). The three main chapters (4–6) are largely based on submitted manuscripts and include the analysis of the syndepositional Upper Rotliegend II tectonic controls of the study area, the comparison to an analogue study site in the Panamint Valley, California, Western United States and the integration of diagenesis and the tectonic phases, which were intensely triggered by halokinesis.

Chapter 4 deals with the combined tectonic-stratigraphic interpretation of three-dimensional seismic reflection, wireline log and core data. Detailed palaeo-topography, fault activity and accommodation analysis through Rotliegend and Zechstein times indicate synsedimentary halfgraben development and fault activity during the Upper Rotliegend II (mainly Elbe Subgroup) deposition. Many Upper Rotliegend structures were reactivated during the Triassic, Jurassic and Cretaceous, often developing enhanced offsets and lateral extent. As a consequence, the original Rotliegend structures and stratigraphy were significantly displaced, ultimately re-arranging the location of former Rotliegend depocentres and associated tight gas reservoir facies.

Research focused on unravelling the structural and sedimentary complexity of the tight gas field requires an integrated approach combining laboratory analysis, numerical modelling and field based analogue studies. Chapter 5 ties directly into the analysis of palaeo-depositional environments using a field analogue study. The study includes a geological framework model for the sedimentary facies distribution strongly influenced from the structural grain. The field analogue-based study provides a detailed model of the sedimentary facies for the tight gas reservoir in Germany prior to the multi-phase tectonic overprint, including the composition and distribution of aeolian sandstone reservoir facies and their relation to fault-induced topography.

The research focus in Chap. 6 lies on the multi-tectonic overprint that leads to salt movement and influences the characteristics of the tight gas field. Taking isopach map analysis, fluid inclusion homogenization temperature measurements and K/Ar ages into account, a distinct correlation between the structural evolution and the timing of diagenetic phases and temperatures has been established.

Chapter 7 focuses on the evaluation of sedimentary facies in core material of the uppermost Upper Rotliegend II in northern central Germany and contains a comparison to the analyzed core material of the north-western German study site.

A final conclusive (Chap. 8) presents the main findings, answers to the questions introduced in this chapter and perspectives.

References

1. BGR (2009) Cramer B, Andruleit H (eds) Energierohstoffe 2009—Reserven, Ressourcen, Verfügbarkeit—Erdöl, Erdgas, Kohle, Kernbrennstoffe, Geothermische Energie, p 117
2. Law BE, Curtis JB (2002) Introduction to unconventional petroleum systems. AAPG Bull 86(11):1851–1852
3. Holditch SA (2006) Tight gas sands. J Petrol Technol 58(6):86–93
4. Holditch SA (2007) Hydraulic fracturing: overview, trends, issues. Drilling Contractor July/ August, pp 116–118
5. Ziegler PA (1982) Geological Atlas of Western and Central Europe: Elsevier Science Ltd., Amesterdam, p 130
6. Legler B (2005) Faziesentwicklung im Südlichen Permbecken in Abhängigkeit von Tektonik, eustatischen Meeresspiegelschwankungen des Proto-Atlantiks und Klimavariabilität (Oberrotliegend, Nordwesteuropa): Schriftenreihe der Deutschen Gesellschaft für Geowissenschaften, vol 47, p 103

Chapter 2
Geological Setting

2.1 Regional Geological Setting

The Middle European Permian/Dyas is subdivided by its facies into the mainly terrestric Rotliegend (302–258 Ma = Million years) and the lagoonal sabkha deposits of the Zechstein (258–251 Ma). The Rotliegend Group/Lower Dyas (Germany), Rotliegend Group (United Kingdom) or Rotliegend Super Group (Netherlands) contains fine grained clastic sediments and evaporites deposited in an arid environment, with a succession of volcanics and evaporites at its base. The Rotliegend contains most of the gas fields of Europe (STD [1]).

In Germany, the Rotliegend is subdivided into: (1) the Lower Rotliegend (including the Altmark Subgroup), which is composed of max. 3,000 m of volcanics erupted from several volcanic complexes. (2) The Upper Rotliegend I (including the Müritz Subgroup), which is due to its only regional extent laterally not traceable. Its biostratigraphic layers match the Upper Rotliegend I in intramontane basins (southern Germany). The lithology is similar to the Upper Rotliegend I intramontane depressions and to the sedimentary layers of the Altmark Subgroup [2]. (3) The up to 2,000 m thick Upper Rotliegend II [3], which provides the framework of this study. It is characterized by continental siliciclastics and minor evaporites, which were deposited under arid to semi-arid climates [4, 5]. The Upper Rotliegend II consists of two subgroups: the Havel and the Elbe Subgroup. In northern central Germany and the German North Sea sector, the Altmark I—IV tectonic pulses, defined by Bachmann and Hoffmann [6], correlate to a series of stratigraphic units that divide the Upper Rotliegend II into four upward-fining successions (Figs. 2.1 and 2.2; [7–10]): the Parchim, Mirow, Dethlingen and Hannover Formations, of which only the upper Dethlingen and Hannover Formations form tight gas reservoir units. These formations have durations of approximately 2 Ma. According to the Stratigraphic Table of Germany (STD 2002) these formations correlate with the late Wordium (264 Ma) to the middle Wuchiapingium (258 Ma) in global stratigraphic context (Figs. 2.1 and 2.2).

A. A. Vackiner, *Sedimentary Facies Reconstruction and Kinematic Restoration of Tight Gas Fields*, Springer Theses, DOI: 10.1007/978-3-642-36047-3_2,

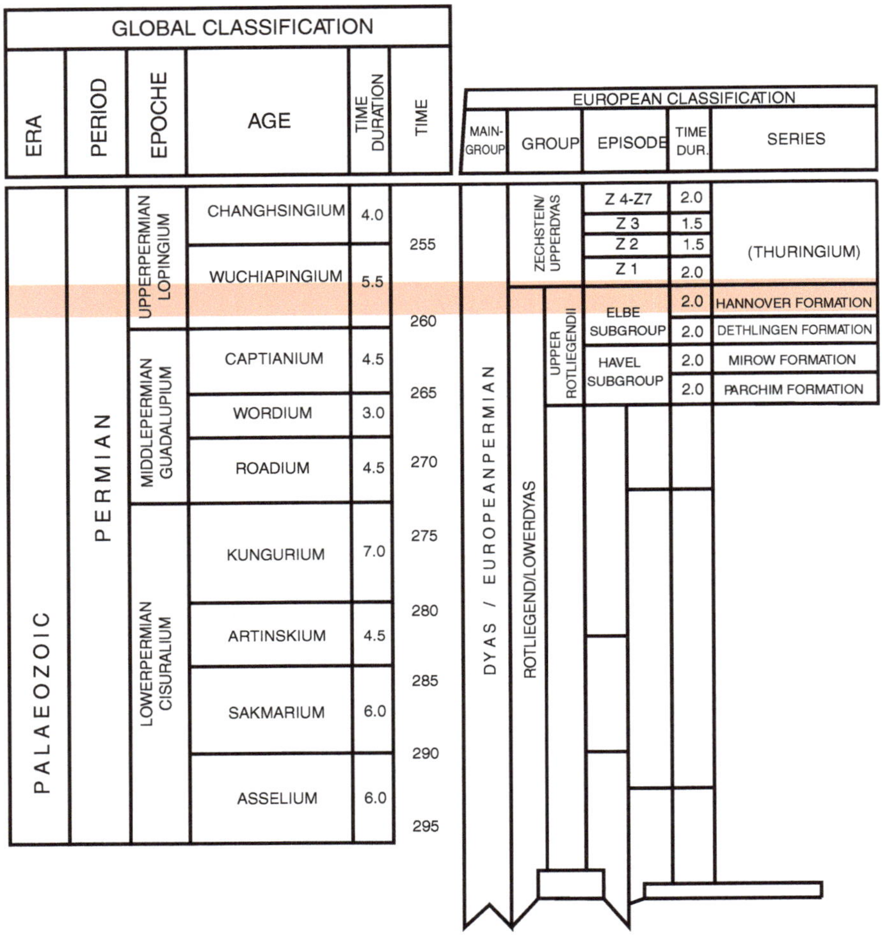

Fig. 2.1 Stratigraphic overview of the Permian; modified from STD (2002). The studied time interval is highlighted in light red

This thesis deals with the Hannover Formation, comprising the Ebstorf, Wustrow, Bahnsen, Dambeck, Niendorf, Munster and Heidberg Members, which were deposited between 260 and 258 Ma (Fig. 2.2).

The study area is located at the south-western margin of the Southern Permian Basin (SPB), which underwent a stepwise enlargement of its depositional area from the centre towards the west and the east [3]. During the Upper Rotliegend II, the SPB had a width (north–south) of 300–600 km and extended over ~1,700 km from the eastern UK to central Poland and the Czech Republic, covering an area of ~430,000 km^2 [11]. The basin was asymmetric and deepest in the north [12, 13]. Today, the basin configuration reflects the cumulative effects of changes of the regional stress regime that affected the re-organization of European crust during

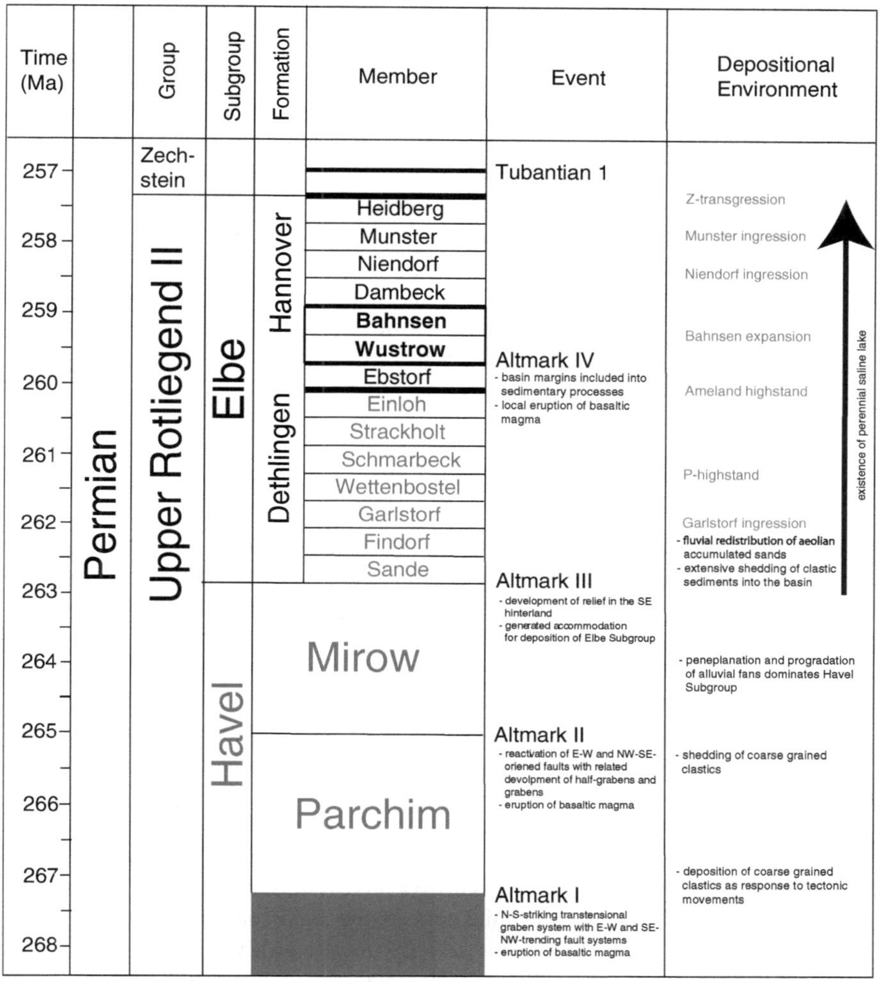

Fig. 2.2 Stratigraphic chart of the *upper* Rotliegend II in northern Germany [2] and contemporaneous tectonic events [28, 6, 14]. The timing of the Elbe Subgroup is based on Gast [7]. Table modified from Stollhofen et al. [3]. Weighted lines mark horizons, which border isopach map thickness (cf. Chap. 4)

decline and after the Variscan Orogeny. The basin comprises several NW–SE-trending pull-apart (en échelon) sub-basins, including the Silverpit/Dutch, North German and Polish Basins, which are separated by north–south and NNW-SSE-trending Variscan basement highs [14].

Since the deposition of the Dethlingen Formation (Figs. 2.1 and 2.2), a perennial saline lake occupied the central part of the SPB [15]. Short-term, but widespread marine transgressions and ingressions into the ephemeral shallow lake are represented by the Garlstorf and Schmarbeck Members of the Dethlingen

Formation and by the Niendorf and Munster Members of the Hannover Formation [16, 17]. These marine incursions required a basinward topographic gradient which resulted from subsidence of the basin floor below sea level, but have also been interpreted to have originated from sea level highstands [18, 19]. Apart from lacustrine and marine deposits, fluvio-aeolian deposits form an important part of the sedimentary record of the Dethlingen and Hannover Formations in the study area. Gast [20] and Rieke et al. [21] concluded that aeolian sediment was supplied by the prevailing easterly winds favouring the accumulation of aeolian dunes on a belt of saline mudflats and sandflats, which fringed the ephemeral shallow lake [22]. These fluvio-aeolian deposits form the major plays targeted by Rotliegend gas exploration [3]. The main source areas of fluvial sediment are located in the Variscan hinterland further south [13, 23].

2.2 Geology of the Study Area

The subsurface study area is located at the eastern boundary of the Ems Graben at the south-western margin of the SPB and is characterized by a mainly N–S-trending Zechstein salt wall situated above an asymmetric Upper Rotliegend II graben. The stratigraphic record of the study area covers the Ebstorf to Heidberg Members (Fig. 2.2). Sedimentary thicknesses range ~200–450 m. The Ems Graben in the central SPB underwent Upper Rotliegend II synsedimentary tectonics. Later subsequent phases of tectonic activity, e.g. rifting in the North Sea during earliest Triassic until late Jurassic to Early Cretaceous [24] overprinted the Rotliegend structural grain.

Source rocks for the gas fields of the Upper Rotliegend II of the study area are Westphalian coals. The top seal is provided by Zechstein evaporites [25]. The reservoir rocks are situated in the Wustrow and Bahnsen Members of the Hannover Formation (Fig. 2.2). The thickness of the sandstone-dominated Wustrow Member is possibly linked to tectonic activity (local basalt volcanism, Soltau high; [2]). Sandstones deposited in the coastal area around the salt lake are of high sedimentary maturity (high poro-perm values, high quartz content) and are therefore of great interest for the gas exploration in the Rotliegend. The Wustrow Member represents the best gas reservoir of the study area [2, 25]. During the Bahnsen Member the Bahnsen ingression took place [19]. The saline lake that occupied the central part of the SPB expanded and wide marginal areas of the basin were influenced by damp to wet synsedimentary conditions [2] resulting in higher clay contents and decreased sedimentary maturity. Until the Zechstein transgression, only minor regression phases with little sand accumulation in basin parallel beach belts occurred [2]. Core data analysis of the study area's Wustrow and Bahnsen Member sediments reveals that they are of fluvio-aeolian origin, including braided stream, aeolian dune and wet to dry interdune deposits. The majority of the aeolian sediment was supplied by prevailing eastern trade winds [20, 21]. In contrast, the source of major fluvial sediment input was located in the

Variscan hinterland towards the south [13, 23]. The preservation of aeolian dunes was governed by tectonic subsidence [26].

The Upper Rotliegend II sediments overlie patchy andesitic to basaltic volcanics (also preserved in the cores) and local Carboniferous highs. Both provided essential local sources for the aeolian sediment supply (e.g. [12, 27]).

The second study area in northern central Germany was investigated with the focus on the sediments overlying the main reservoir successions. It is situated in the southern part of the SPB. During the Dambeck Member of the Hannover Formation thin sandstones accumulated, but sabkha and lake sediments prevailed [2]. The area was influenced by short-term, widespread marine ingressions (Niendorf and Munster Members of the Hannover Formation; [16, 17]) into the shallow lake that occupied the central SPB since the onset of the Dethlingen Formation.

References

1. Menning M, Hendrich A (eds) and Deutsche Stratigraphische Kommission (2002) Stratigraphische Tabelle von Deutschland 2002. Tafel 96 × 130 cm oder Falt-Tafel A4; Potsdam (GeoForschungsZentrum), Frankfurt a. M. (Forsch.-Inst. Senckenberg). ISBN 3-00-010197-7
2. Plein E (1995) Stratigraphie von Deutschland I, Courier Forschungsinstitut Senckenberg, vol 183
3. Stollhofen H, Bachmann NGH, Barnasch J, Bayer U, Beutler G, Franz M, Kästner M, Legler B, Mutterlose J, Radies D (2008) Upper Rotliegend to early cretaceous basin development. In: Littke R, Bayer U, Gajewski D, Nelskamp S (eds) Dynamics of complex intracontinental basins; the central European basin system. Springer, Berlin, pp 181–210
4. Glennie KW (1972) Permian Rotliegendes of northwest Europe interpreted in light of modern desert sedimentation studies. AAPG Bull 56:1048–1071
5. Glennie KW (1983) Early Permian (Rotliegendes) palaeowinds of the North Sea. Basin analysis and sedimentary facies; sedimentology at various scales. Sediment Geol 34(2–3) 245–265
6. Bachmann GH, Hoffmann N (1997) Development of the Rotliegend basin in Northern Germany. In: Geologisches Jahrbuch RD Mineralogie, Petrographie, Geochemie, Lagerstaettenkunde, vol 103. pp 9–31
7. Gast R (1995) Sequenzstratigraphie. In: Plein E (ed) Stratigraphie von Deutschland I; Norddeutsches Rotliegendbecken—Rotliegend-Monographie Teil II. Courier Forschungsinstitut Senckenberg, vol 183. pp 47–54
8. Gebhardt U, Schneider J, Hoffmann N (1991) Modelle zur Stratigraphie und Beckenentwicklung im Rotliegenden der Norddeutschen Senke. Geol Jahrb A127:405–427
9. Hoffmann N, Kamps H-J, Schneider J (1989) Neuerkenntnisse zur Biostratigraphie und Paläodynamik des Perms in der Nordostdeutschen Senke—ein Diskussionsbeitrag. Z Angew Geol 35:198–207
10. Hoffmann N (1990) Zur paläodynamischen Entwicklung des Präzechsteins in der Nordostdeutschen Senke. Niedersächsische Akademie der Wissenschaften, Geowissenschaftliche Veröffentlichung, vol 4. pp 5–18
11. van Wees JD, Stephenson RA, Ziegler PA, Bayer U, McCann T, Dadlez R, Gaupp R, Narkievicz M, Bitzler F, Scheck M (2000) On the origin of the southern Permian basin, central Europe. Mar Pet Geol 17:43–59

12. McCann T (1998) The Rotliegend of the NE German basin: background and prospectivity. Pet Geosci 4:17–27
13. Plein E (1993) Bemerkungen zum Ablauf der palaeogeographischen Entwicklung im Stefan und Rotliegend des Norddeutschen Beckens. Observations on Stephanian and Rotliegendes palaeogeography in the North German Basin. In: Zur Geologie und Kohlenwasserstoff-Fuehrung des Perm im Ostteil der Norddeutschen Senke. Geology and hydrocarbon potential of the Permian rocks of the eastern North German Basin: Geologisches Jahrbuch. Reihe A: Allgemeine und Regionale Geologie BR Deutschland und Nachbargebiete, Tektonik, Stratigraphie, Palaeontologie, vol 131. pp 99–116
14. Börmann C, Gast R, Görisch F (2006) Structural and sedimentological analysis of an early late Rotliegendes graben based on 3D seismic and well log data, German North Sea. Pet Geosci 12:195–204
15. Gast R, Gaupp R (1991) The sedimentary record of the late Permian saline lake in N.W. Germany. In: Renaut RW, Last WM (eds) Sedimentary and paleolimnological records of saline lakes. National Hydrology Research Institute, Saskatoon, pp 75–86
16. Gast R, Gebhardt U (1995) Elbe Subgruppe. In: Plein E (ed) Stratigraphie von Deutschland I; Norddeutsches Rotliegendbecken—Rotliegend-Monographie Teil II. Courier Forschungsinstitut Senckenberg, vol 183. pp 121–145
17. Legler B, Gebhardt U, Schneider JW (2005) Late Permian non-marine—marine transitional profiles in the central southern Permian basin. Int J Earth Sci 94:851–862
18. Stemmerik L, Ineson JR, Mitchell JG (2000) Stratigraphy of the Rotliegend group in the Danish part of the northern Permian basin, North Sea. J Geol Soc 157:1127–1136
19. Stemmerik L (2001) Sequence stratigraphy of a low productivity carbonate platform succession: the upper Permian Wegener Halvø formation, Karstryggen area, East Greenland. Sedimentology 48:79–97
20. Gast RE (1988) Rifting im Rotliegenden Niedersachsens, Rifting in the Rotliegendes of Lower Saxony: Die Geowissenschaften Weinheim, vol 6, no 4. pp 115–122
21. Rieke H, Kossow D, McCann T, Krawczyk C (2001) Tectono-sedimentary evolution of the northernmost margin of the NE German basin between uppermost Carboniferous and late Permian (Rotliegend). Geol J 36(1):19–38
22. Gast RE (1991) The perennial Rotliegend saline lake in NW Germany. Geol Jb A119:25–59
23. Glennie KW (1990) Introduction to the petroleum geology of the North Sea, vol 3. Wiley-Blackwell, p 416
24. Ziegler PA (1990) Geological atlas of western and central Europe. Shell, 2nd edn. The Hague, p 239
25. Schwarzer D, Littke R (2007) Petroleum generation and migration in the, 'tight gas' area of the Germany Rotliegend natural gas play: a basin modelling study. Pet Geosci 13:37–62
26. Kocurek G (2003) Limits on extreme eolian systems; Sahara of Mauritania and jurassic navajo sandstone examples. In: Chan MA, Archer AW (eds) Extreme depositional environments; mega end members in geologic time: geological society of America special paper, vol 370. pp 43–52
27. Glennie KW (1990) Rotliegend sediment distribution; a result of Late Carboniferous movements. In: Hardman RFP, Brooks J (eds) Proceedings of tectonic events responsible for Britain's oil and gas reserves, vol 55. Geological Society of London, Special Publications, pp 127–138
28. George GT, Berry JK (1997) Permian (upper Rotliegend) synsedimentary tectonics, basin development and palaeogeography of the southern North Sea. In: Ziegler K, Turner P, Daines SR (eds) Petroleum geology of the southern North Sea; future potential, vol 123. Geological Society of London, Special Publications, pp 31–61

Chapter 3
Data and Methods

3.1 Data and Methods

The study is based on multi-disciplinary data, comprising 3D seismic, wireline and core data. Seismic data include a 293 km^2 3D-seismic volume in pre-stack depth-migration (PSDM) and post-stack time-migration (PSTM). Furthermore, the seismic data set contains a regional, 100 km long, W-E-oriented, PSTM 2D seismic line that crosses the southern part of the 3D seismic survey (Fig. 3.1). The regional 2D seismic line was depth-converted using a velocity model, based on sonic log calibration. Data from 14 wells, including digital wireline logs from seven wells, core data from four of these wells, FMI/FMS logs from one of these wells, well reports and logs on paper from the other seven wells were also considered in this study. Six wells are located on a structural high in the eastern part of the survey, the area with the best data coverage. An overview of the main data set is displayed in Figs. 3.1 and 3.2.

In addition to the data set from the key study area in north-western Germany, a second data set includes core material and well records of 8 wells and a 2D PSTM seismic line (location sketched in Fig. 1.2). The study site is further refered to as northern central German study area. The seismic line was scanned and horizon quality was improved by image editing. The interpretation was depth-converted on the basis of a velocity model based on well records from several wells. Well control is available from Well I, which is located in the central part of the 2D seismic line.

The structural interpretation of the seismic data was mainly carried out on the depth migrated seismic data. In order to link the regional 2D seismic line to the 3D seismic cube a depth-conversion was carried out on the basis of interval velocities derived from sonic velocities. The seismic interpretation and the depth-conversion were mainly accomplished using Schlumberger's Petrel software package (Fig. 3.2).

For the analysis of accommodation and fault-induced palaeo-topography, isopach maps were generated. Isopachs are lines connecting points of equal true stratigraphic thickness. For the compilation of isopach maps the thickness of a

A. A. Vackiner, *Sedimentary Facies Reconstruction and Kinematic Restoration of Tight Gas Fields*, Springer Theses, DOI: 10.1007/978-3-642-36047-3_3,

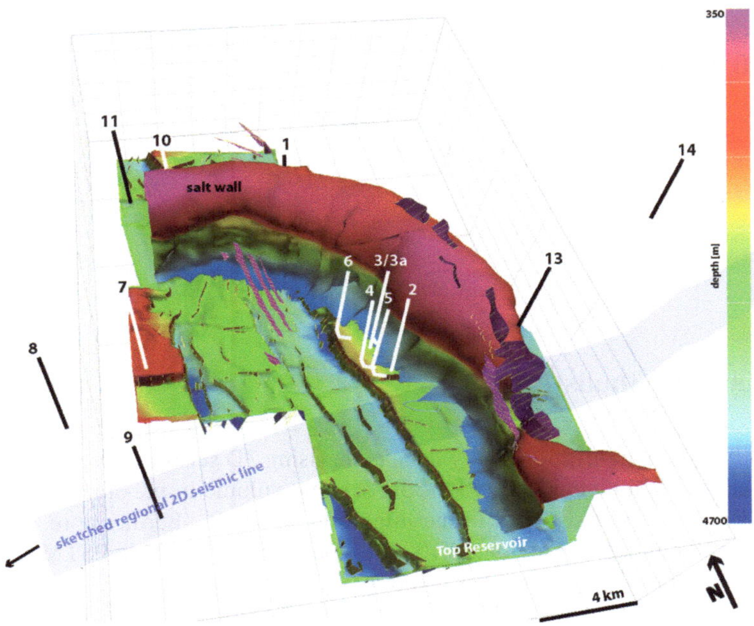

Fig. 3.1 Overview of the seismic data set from north–western Germany, view from the south. 3D seismic survey displayed with faults, salt wall and Top Reservoir depth map. Faults of Zechstein overburden strata are displayed in pink. The regional 2D seismic line (blue shade) crosses the 3D seismic survey in the south and continues eastwards for another ~80 km. The existing wells are depicted in black and white

certain stratigraphic interval was measured perpendicular to the upper and the lower horizon of the interval. Enlarged sedimentary thicknesses indicate sedimentary depocentres such as small sub-basins and grabens. Hangingwall fault-induced depocentres develop during synsedimentary fault activity. For the detailed accommodation analysis of the Permian stratigraphic interval, seven horizons were interpreted over the 3D depth-migrated data set. Permian marker horizons were from top to base: Top Zechstein, Top A2 Zechstein, Top Z1 Zechstein, Top Rotliegend, Top Bahnsen as top of the gas reservoir (Fig. 3.1), Top Ebstorf as base of the gas reservoir, and base of the Upper Rotliegend II. Furthermore, 93 NNW-SSE and north–south-trending normal faults and 30 west–east-trending faults, controlling the Upper Rotliegend II interval, were interpreted on the basis of the 3D data; 18 additional Rotliegend faults were observed on the regional 2D line. Isopach maps were generated for the particular stratigraphic units bounded by seismic reflectors within the reservoir interval.

In addition to the interpretation of the Permian interval in reservoir vicinity, the tectonic development of the study area through time was evaluated. 13 isopach maps of key stratigraphic units were calculated from interpreted seismic horizons: Rotliegend, Zechstein, Lower Buntsandstein, Upper Buntsandstein (Röt Member),

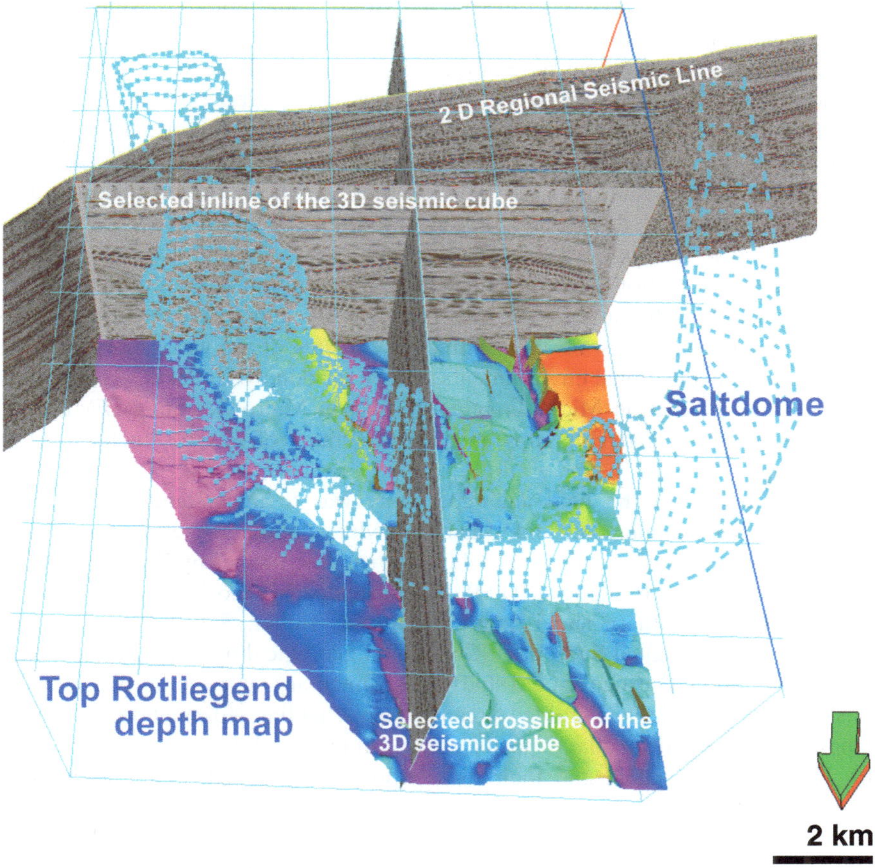

Fig. 3.2 Overview of the seismic data set from north–western Germany, view from the north. 3D seismic survey displayed with structural model, salt diapir extent and Top Rotliegend depth map. The regional 2D seismic line crosses the 3D seismic survey in the south and continues eastwards for another ~80 km. The salt dome extent is extrapolated towards the west. Shape of the salt dome outside the seismic survey based on Jaritz [16] and Baldschuhn et al. [17]

Lower Muschelkalk, Middle Muschelkalk, Upper Muschelkalk, Lower Keuper, Upper Keuper ± Jurassic, Lower Cretaceous, Cenomanian-Santonian (early Upper Cretaceous), Campan (Upper Cretaceous) and Maastrichtian (latest Upper Cretaceous). Differential sediment loads as response to regional and local tectonics and to salt kinematics provide an important input parameter for unravelling the tectonic history. Areas of enhanced or reduced sediment thickness indicate tectonic activity, which is related to salt movement in the study area. Enhanced sediment thickness on the hangingwall of normal faults in comparison to less sediment thickness on the footwall characterizes the existence of fault-induced palaeo-relief.

In addition, the following 10 interpreted stratigraphic horizons were used for stepwise sequential retro-deformation: Top Rotliegend, Top Zechstein, Top Solling (Upper Buntsandstein; as base of the Röt evaporites), Base Muschelkalk, Base Keuper, Base Upper Keuper, Base Cretaceous, Base Upper Cretaceous, Base Maastrichtian (latest Upper Cretaceous) and Base Tertiary. The primary objective of the sequential restoration is to calculate and remove the effects of processes that have influenced and controlled the evolving basin geometry during specified time (e.g. [1]). A detailed description of the algorithms used for retro-deformation purposes is described in Sect. 7.2.

For the determination of sedimentary facies distribution well information from 14 wells including digital wireline logs from 7 wells and core data from 4 wells were used. The analysis of the sedimentary facies from well cores and the description of depositional environments will be introduced in Sect. 3.1. Dipmeter logs and one Formation Micro Imaging/Formation Micro Scanning (FMI/FMS) log were used for analyzing dips and dip directions of foresets in the aeolian successions.

The main aim of the field analogue study was to understand the reservoir architecture and its' elements and to gain knowledge about properties and dimensions of the reservoir rocks. Besides field work, satellite images by Google Earth™, Light Detection And Ranging (LIDAR) data provided by GeoEarthScope (Project: GeoEarthScope Southern and Eastern California, Target: SoCal_Panamint) and United States Geological Survey (USGS) digital geological maps [2, 3] were taken into consideration. During the field study the sedimentary facies comprising alluvial fans of different angles and braided stream systems, aeolian sediment bodies such as dunes, interdune and sandflat deposits, and the mudflat sediments of the ephemeral dry lake, were classified and their distribution was mapped. Fault orientations were adapted from Jennings [2] and Jennings et al. [3] and compared to the measurements from high-resolution LIDAR data, field observations, and satellite images. The composition of sediment samples from dune sands and clays of the dry lake surface collected during the field study were analyzed by X-ray diffractometry (XRD) and on grain compound thin-sections. The results of sediment analyses were used to determine the origin and compositional maturity of especially the aeolian sediments.

3.2 Methods: Sedimentary Facies Analysis from Core Material

Facies interpretation from core analysis is one of the key methods applied in this doctoral thesis. Detailed sedimentary facies subdivision was possible because of the very large footage of Upper Rotliegend II core ($\sim$600 m) examined during the study. Core material is available from 3 wells (Wells 2, 3 and 3a) of the northwestern German study area (Fig. 3.1) and from 4 wells of the northern central

German study area (Wells I–IV; Fig. 7.2). The analysis provides crucial information about the palaeo-depositional environment during the sedimentation of today's tight gas reservoir rocks. Core material was analyzed focusing on grain sizes, clay content, sedimentary structures and finally it was assigned to sedimentary facies.

In the following, the different fluvio-aeolian sedimentary facies associations, which were reconstructed from the main north-western and from the second northern central German study areas, are introduced and characteristic examples displayed. Lateral attached and vertical stacked sedimentary facies of ephemeral ponds/lakes, pond/lake margins, aeolian mudflats, wet, damp and dry sandflats, aeolian dunes and fluvial facies associations occur in typical playa basins [4, 5; Fig. 3.3].

N C GER in the core pictures' figure captions indicates, that the core material shown has its origin in the northern central German study site, whereas NW GER is short for the origin of core material from the north-western German tight gas field.

3.2.1 Pond/Lake

Pond or lake deposits consist of 100 % clay. They are structure-less to laminated red (Fig. 3.4a–c), minor green to grey sediments. Pond or lake deposits occur in areas of ephemeral dry lakes (Fig. 3.5a) and in interdune areas (Fig. 3.5b), in which the (saline) water table intersects the interdune troughs [7]. In some areas, temporary playa lakes may form in interdune depressions following extensive rainfall [8]. In many cases, salt efflorescence (Fig. 3.4c) can be recognized along horizontal, bedding parallel, and minor along vertical small uncemented fracture systems (Fig. 3.4b).

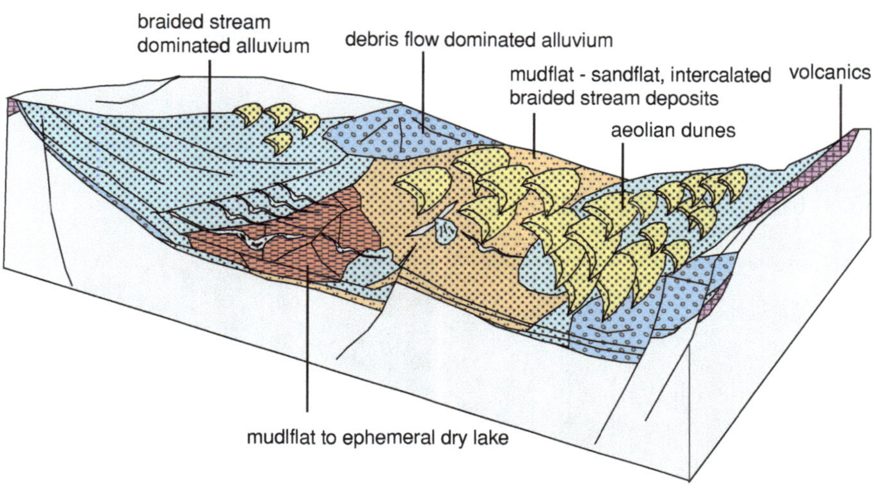

Fig. 3.3 Diagram illustrates depositional environments in a typical playa basin; modified from [4, 5] and [8]

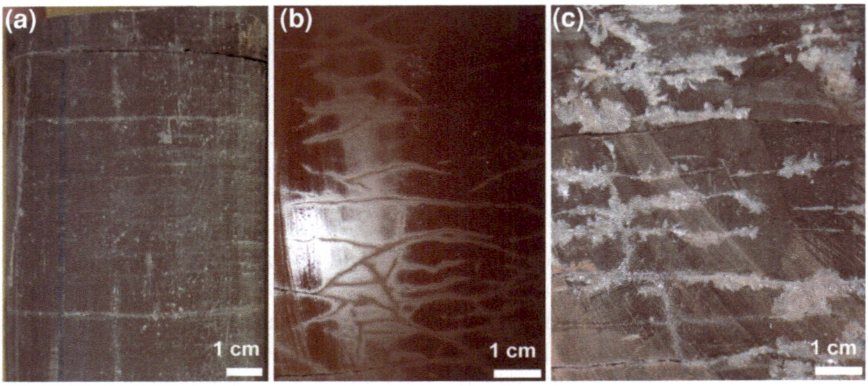

Fig. 3.4 **a** Pond deposit with up to 1 mm thick sand layers. NW GER, Well 3. **b** Pond deposit with vertical and minor horizontal micro-fracture network. N C GER, Well 4. **c** Pond deposit with vertical and minor horizontal micro-fracture network. The core shows salt effloresces along fracture network. N C GER, Well II

3.2.2 Pond/Lake Margin

Pond/lake margin deposition is characterized by irregular cm to dm scale intercalations of fine to medium grained sandstones and clay. Ripple lamination is common. Typically, lake margin sedimentation takes place in a sub-aquatic milieu. Aeolian sandstones are intercalated as a result of prograding dune sets into lakes/ponds (e.g. [4, 5]; Figs. 3.6a, b).

An alternative model concerning a depositional environment that causes close irregular intercalations of clay- and sandstone is a fluvial origin. In that case, the sandstones are the result of fluvial channels respectively sheetfloods discharging into a lake environment.

Fig. 3.5 **a** Shallow sabkha lake, Panamint Valley, CA. **b** Dune migrates into interdune pond, Death Valley, CA

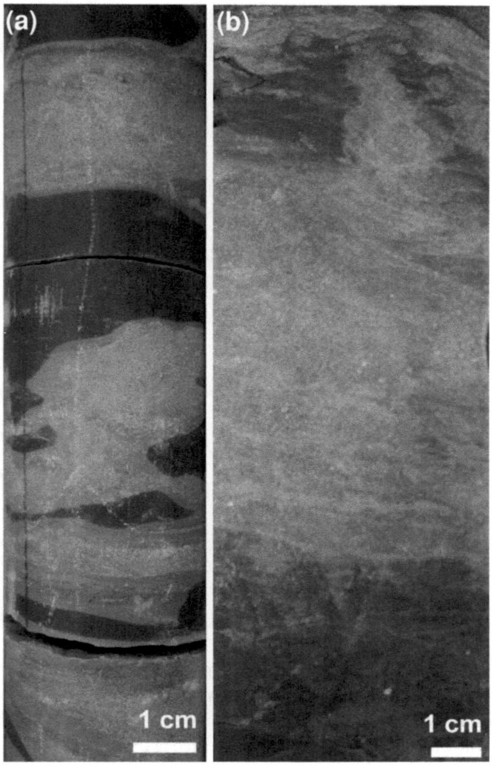

Fig. 3.6 **a** Lake margin deposit. NW GER, Well 3. **b** Lake margin deposit with post-sedimentary sand injection. N C GER, Well I

3.2.3 Aeolian Mudflat

Aeolian mudflat deposits are wavy laminated claystones with intercalated lenses of siltstone and sandstone (Figs. 3.7a–d, 3.8). They are characterized by a clay content of >50 % [6]. The deposition takes place in a mainly sub-aquatic milieu without recognizable currents, most likely in ephemeral ponds or lakes. Desiccation cracks and soft clasts indicate periodical subaerial exposition of the mudflat deposits, during which adhesive sands and silts are synsedimentarily blown onto the damp sediment surface [8]. Furthermore, adhesive ripples occur as characteristic sedimentary structures [7]. In most instances, the mudflat deposits are not laminated, but contain abundant post-sedimentary dewatering and most likely desalinization structures, such as convolute bedding, flame structures and sand dykes (esp. Fig. 3.7d).

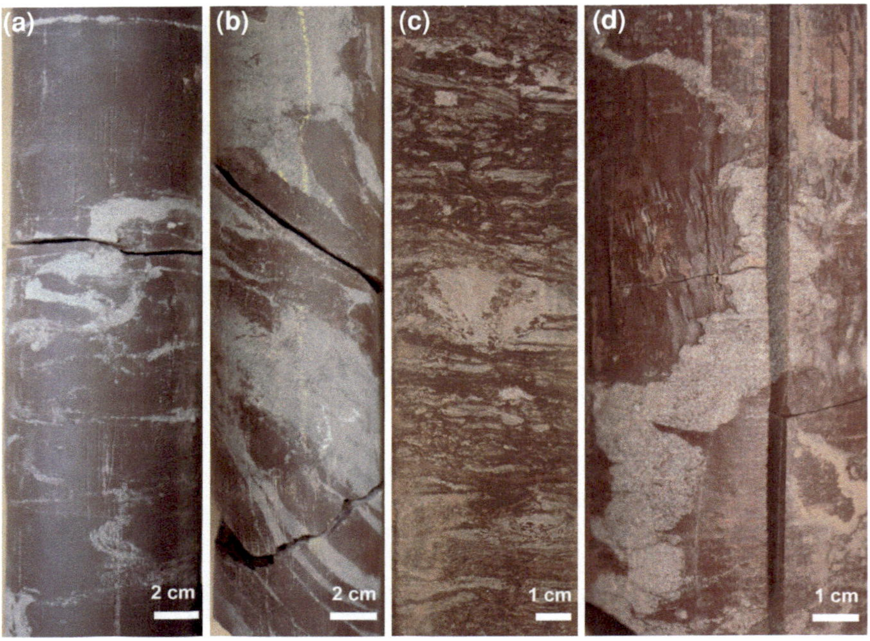

Fig. 3.7 **a** Aeolian mudflat deposit with post-sedimentary convolute bedding. NW GER, Well 3. **b** Aeolian mudflat deposit. NW GER, Well 3a. **c** Aeolian mudflat deposit with post-sedimentary convolute bedding. N C GER, WellII. **d** Aeolian mudflat with sand injection. Core picture from Antrett [18]

Fig. 3.8 Interdune aeolian mudflat to wet sandflat. Picture taken in Panamint Valley, CA

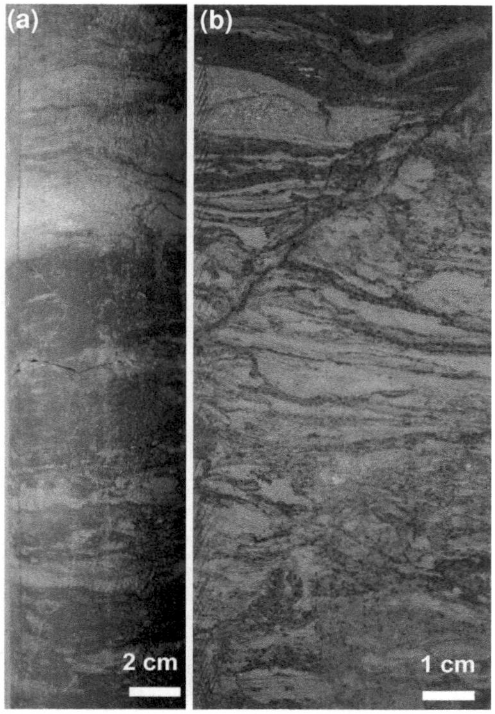

Fig. 3.9 **a** Wet sandflat deposit. NW GER, Well3. **b** wet sandflat deposit. N C GER, Well II

3.2.4 Wet to Damp Sandflat

Wet sandflat deposits are very fine to fine grained, poorly sorted sandstones and siltstones with clay contents of 20–50 % ([6]; Fig. 3.9a–b). Damp sandflat deposits are fine grained to medium grained sandstones with minor siltstones. Clay contents are <20 % ([6]; Fig. 3.10a–b). Both are characterized by the occurrence of discontinuous, irregular to wavy argillaceous adhesion structures (e.g. ripples) and small scale contortions (<0.2 m amplitude). They are often accompanied by irregular, lensoid to aligned concentrations of sand [9, 10] with abundant post-sedimentary dewatering structures. The primary sedimentary layering is disturbed by precipitation and dissolution of salt below the sediment surface. In some cases, the occurrence of grain flow deposits indicates lateral connectivity to dry aeolian depositional areas. Adhesive sands and deflation lags are common. Wet and damp sandflats occur as interdune or dune field marginal units and develop due to deposition on a damp substrate [9], overlying a shallow groundwater table. For the sedimentation of wet sandflat deposits the surface is supposed to be periodically flooded.

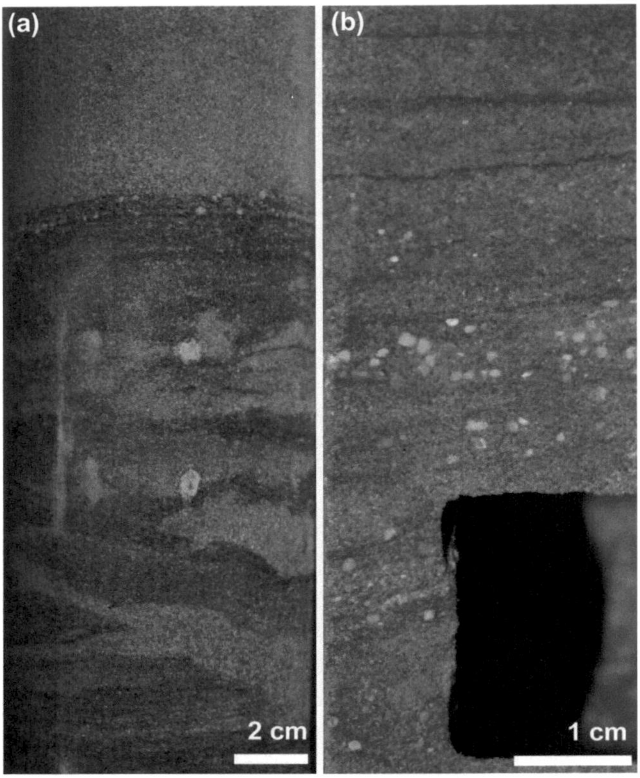

Fig. 3.10 **a** Damp sandflat deposit truncated by deflation lag. NW GER, Well 3. **b** Damp sandflat deposit with adhesion lags. N C GER, Well IV

3.2.5 Dry Sandflat

Dry sandflat deposits are characterized by mature, (very) well bedded to homogeneous, fine to coarse grained sandstones. In most cases they show typical aeolian bimodal grain size distribution and low-angle translatent wind-ripple strata ([9]; dip angles of less than 5°; Fig. 3.11). Small dunes and ripple remnants (Fig. 3.12a), granule ridges, lag deposits and internal low-angle erosion surfaces are common [8]. Dry sandflats might be part of a dune set or fringe a dune accumulation (Fig. 3.12a) and also occur isolated in areas of high wind velocities. They originate from a fine sand supply, a shallow groundwater table or surface stabilization (Fig. 3.12b; [8]). Sedimentation in dry interdune environments generally occurs relatively slowly as a result of vertical rather than lateral accretion, with periods of deflation.

Large thicknesses of dry sandflats only occur in sand sheets that develop on the up-wind (proximal) or down-wind (distal) margins of the dune fields during extended periods of strong wind and/or when sand supplies are limited [10–12, 14].

Fig. 3.11 Dry sandflat deposits with coarse grained grain flows. N C GER, WellII

3.2.6 Aeolian Dune (base)

Aeolian dune deposits are (very) well bedded, fine to coarse grained, mature sandstones (Figs. 3.13a–b, 3.14). They are characterized by semi-tabular (large-scale) cross-stratifications with foresets that show bimodal grain size lamination. Coarser grained grain flow deposits are closely intercalated with finer grained suspension fall out deposits. Grain sizes may vary depending on the prevailing wind velocities, the source area and the location of dunes in the dune field. Aeolian dunes are typically characterized by cross-stratifications with upward increasing dip angles. Aeolian dune bases show dip angles of 5–15°, whereas aeolian dunes show dip angles of 15–35° at an average of $\sim$22° ([10]; Fig. 3.15a).

In this study, dune deposits were subdivided into barchanoid (Fig. 3.13b) and aklé dunes (small amalgamated dune sets; Figs. 3.13a, 3.15b). Barchanoid dunes (Fig. 3.13b) are crescent-shaped dunes with their wings tapering down-wind [8]. Individual dunes are separated from neighbouring dunes by either solid rock floor or immobile coarse pebbles (deflation lags; [8]). The occurrence of barchanoid dunes indicates limited sand supply and/or high wind velocities [13, 15]. Due to the high dune ridge sinuosity in barchanoid dunes, the sandstones show a considerable spread in palaeo-transport indicators. The individual layers truncate each other. In comparison, aklé dunes (Fig. 3.13a) are sinuous-crested dune types, which developed under plentiful sand supply. They show large sets of internal cross-stratification with unimodal dip direction [8].

Fig. 3.12 **a** Dry sandflat in marginal position of the dune field, Panamint Valley, CA. **b** Damp sandflatin in marginal position of the dune field, Panamint Valley, CA

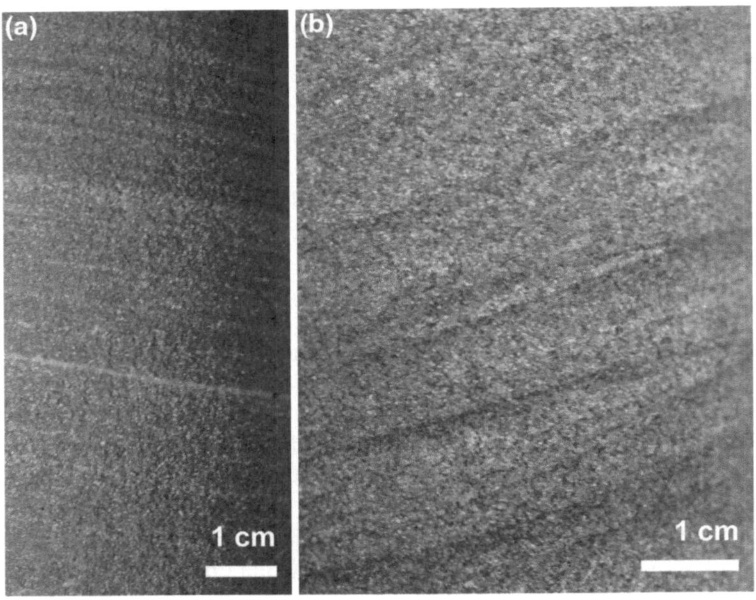

Fig. 3.13 **a** Aeolian aklé dune base deposit. NW GER, Well 3. **b** Aeolian barchanoid dune deposit. NW GER, Well 3

3.2.7 Low Energetic Meandering Fluvial Deposits

Grain sizes vary between clay (predominantely) and coarse grained sand. Small-scale tangential epsilon (ε)-cross-stratification is common. The lower part of the small scale cross-beds is characterized by fine grained sediments, the upper part by comparative coarser grained sediments. The fluvially influenced respectively reworked sediments are further characterized by small rip-up clasts and clay cullets (Fig. 3.16).

Fig. 3.14 Aeolian dune base and dune deposits. N C GER, Well II

Fig. 3.15 **a** Dune field, Eureka Valley, CA. **b** Erosional unconformities and foresets in aeolian dune after heavy rainfall, Panamint Valley, CA

3.2.8 High Energetic Braided Stream Fluvial Deposits

Braided stream systems (Fig. 3.17) are characterized by very coarse grained (conglomeratic) gravel bar deposits and finer grained fluvial channel sands with cross-stratifications (Fig. 3.18). The gravel bar conglomerates contain extraformational clasts and gravels or a mixture of intra- and extraformational clasts and gravels. Internally, the conglomerates may display crude bedding, pebble alignment or imbrication, normal or reverse grading, or they may be completely disorganized [10]. Sandstone interbeds are stratified and contain floating clasts. This proximal fluvial facies can pass upward into other fluvial facies or be abruptly overlain by aeolian facies [10].

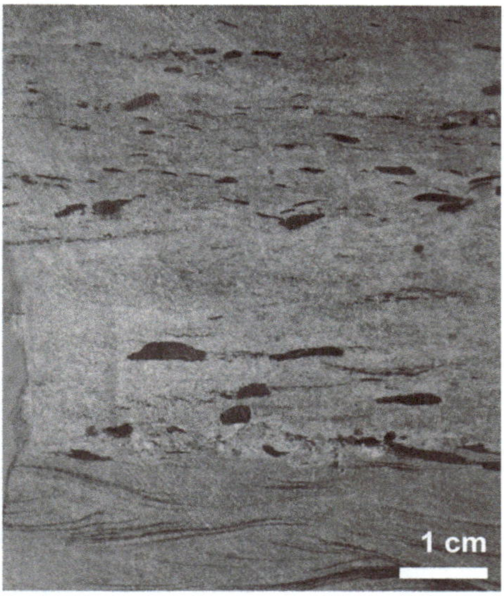

Figs. 3.16 Low energetic fluvial deposits with ε -cross-stratification and clay rip-up clasts (right). N C GER, Well I

Fig. 3.17 Shallow sand-filled fluvial channel of a dry braided stream fluvial system, Panamint Valley, CA

Fig. 3.18 Braided stream deposit. NW GER, Well 3

References

1. Rowan MG (1993) A systematic technique for the sequential restoration of salt structures. In: New insights into salt tectonics; collection of invited papers reflecting the recent developments in the field of salt tectonics, Cobbold. Tectonophysics, vol 228(3–4), p 331–348
2. Jennings CW (1975) Fault map of California with location of volcanoes, thermal springs, and thermal wells: California Division of Mines and Geology Geologic Data Map No. 1, scale 1:750,000, 1 sheet
3. Jennings CW, Bryant WA, Saucedo G (2010) Fault activity map of California. California geological survey 150th anniversary: California geologic data map series map no 6, scale 1:750,000, 1 sheet
4. Hardie LA, Smoot JP,Eugster HP (1978) Saline lakes and their deposits: a sedimentological approach. In: Matter A, Tucker ME (eds) Modern and Ancient Lake Sediments. International Association of Sedimentologists Special Publication, vol 2, p 7–42
5. Rosen MR (1994) The importance of groundwater in playas: a review of playa classification and the sedimentology and hydrology of playas. In: Rosen MR (ed) Paleoclimate and basin evolution of playa systems. Geological Society of America, Special Paper, vol 289, p 1–18
6. Amthor JE, Okkerman J (1998) Influence of early diagenesis on reservoir quality of rotliegende sandstone, Northern Netherlands. AAPG Bull 82:2246–2265
7. Glennie KW (1970) Desert sedimentary environments. Elsevier, Developments in Sedimentology, vol 14, Amsterdam, p 222
8. Leeder M (1999) Sedimentology of sedimentary basins—from turbulence to tectonics. Blackwell Science, Oxford, p 608
9. Mountney NP, Jagger A (2004) Stratigraphic evolution of an erg margin aeolian system: the Permian Cedar Mesa Sandstone, SE Utah, USA. Sedimentology 51:713–743
10. George GT, Berry JK (1993) A new palaeogeographic and depositional model for the Upper Rotliegend of the UK sector of the Southern North Sea. In: North CP, Prosser DJ (eds) Characterization of Fluvial and Aeolian Reservoirs. Geological Society of London, Special Publication, vol 73, pp 291–319
11. Fryberger SG, Ahlbrand TS, Andrews S (1979) Origin, sedimentary features and significance of low-angle eolian 'sand sheet' deposits. Great Sand Dunes National Monument and vicinity, Colorado. J Sed Petrol 49:733–746
12. Kocurek G, Nielson J (1986) Conditions favourable for the formation of warm-climate aeolian and sand sheets. Sedimentology 33:751–816
13. Kocurek G, Townsley M, Yeh E, Havholm KG, Sweet ML (1992) Dune and dune field development on Padre Island, Texas, with implications for interdune deposition and water-table-controlled accumulation. J Sed Petrol 62:622–635
14. Lancaster N (1995) Geomorphology of Desert Dunes. Routledge, London, p 312
15. Bagnold RA (1954) The Physics of blown Sands and desert dunes, 2nd edn. Chapman and Hall, London 265p
16. Jaritz W (1973) Zur Entstehung der Salzstrukturen Nordwestdeutschlands. Geol Jb A 10:1–77
17. Baldschuhn R, Best G, Binot S, Brückner-Röhling S, Deneke E, Frisch U, Hoffmann N, Jürgens U, Kockel F, Krull P, Röhling H-G, Sattler-Kosinowski S, Stancu-Kristoff G, Zirngast M (1999) Geotektonischer Atlas von Nordwest-Deutschland und dem deutschen Nordsee-Sektor. Tectonic Atlas of Northwest Germany and the German North Sea Sector. In: Baldschuhn R, Binot F, Frisch U, Kockel F (eds) Geologisches Jahrbuch Reihe A, vol 153, p 88, 3CDs
18. Antrett P (2011) Characterization of an upper permian tight gas reservoir—a multidisciplinary, multi-scale analysis from the Rotliegend. Northern Germany, Dissertation, p 125

Chapter 4
Syndepositional Tectonic Controls and Palaeo-Topography of a Permian Tight Gas Reservoir in NW Germany

4.1 Introduction

In recent years, tight gas reservoirs have received increased exploration interest as new technological developments such as horizontal drilling and hydraulic fracturing simplify tight gas exploitation. The term ''tight gas reservoir'' is not uniformly defined. Law and Curtis [1] defined low permeability sandstone reservoirs as those exhibiting permeabilities <0.1 mD. This value has been committed since as a political margin in the United States to determine which wells deserve governmental funding for gas production [2]. In contrast, the German Society for Petroleum and Coal Science and Technology (DGMK) refers to a tight gas sandstone reservoir if permeabilities are <0.6 mD.

Controls on tight gas reservoir quality include reservoir architecture, fluid dynamics, cement types, and pore structure, parameters, which have seldom been analyzed in integrated studies to date. In the study area, the tight nature of the gas reservoir has been mainly attributed to the presence of quartz overgrowths and fibrous illite, which reduced sandstone permeabilities by cementing pore spaces and pore throats [3, 4]. An additional factor controlling reservoir quality, however, can be syn- and post-sedimentary fault activity, which may have a significant impact on fluid flow due to changes in the dynamic behaviour of faults and fracture zones [5–7]. Faults do not only influence diagenesis, mineralization and fluid flow that are favoured in zones of enhanced hydraulic activity, but can also control sedimentary thicknesses and facies distributions across them [8]. Synsedimentary fault activity causing e.g. footwall highs potentially contributes to the development of suitable reservoir rocks such as braided-stream dominated alluvial fans or up-wind/down-wind accumulations of aeolian sand bodies [9]. The latter are among the highest quality Rotliegend reservoir rocks due to their high textural and compositional maturity.

This study focuses on a tight gas field in NW Germany, located in East Frisia, east of the Ems Graben and the Groningen High (Fig. 1.2). Reservoir rocks in this field are of fluvio-aeolian origin, including barchanoid dune, wet to dry interdune, braided fluvial and alluvial fan facies that are assigned to the Upper Rotliegend II

A. A. Vackiner, *Sedimentary Facies Reconstruction and Kinematic Restoration of Tight Gas Fields*, Springer Theses, DOI: 10.1007/978-3-642-36047-3_4,

interval (Fig. 2.2; [10–14, 38]). In the study area, the Upper Rotliegend II unit thickness ranges between 180 and 450 m, mainly comprising the Ebstorf to Heidberg Members of the Upper Rotliegend II Elbe Subgroup (Fig. 2.2). It will be shown that fault-related palaeo-topography and fault-related depressions may control trapping of reservoir rock sandbodies of various depositional origin. Such an analysis provides crucial information for undrilled areas and areas that underwent multi-phase tectonics. The documentation of fault activity through time and the analysis of the influence of active faulting on the sediment distribution are thus considered to be an important element in the successful prediction of gas reservoir rock locations.

4.1.1 Geological Framework

The study area is located at the SW margin of the WNW–ESE-trending Southern Permian Basin (Fig. 1.2). During the Upper Rotliegend II, the Southern Permian Basin (SPB) had a width (north–south) of 300–600 km and extended over ~1,700 km from the eastern United Kingdom (UK) to central Poland and the Czech Republic, covering an area of ~430,000 km^2 [15]. The basin was asymmetric and deepest in the north [16, 17]. Today, the basin configuration reflects the cumulative effects of changes of the regional stress regime that affected the re-organisation of European crust during decline and after the Variscan Orogeny. The basin comprises several NW–SE-trending pull-apart (en échelon) sub-basins including the Silverpit/Dutch, North German and Polish Basins, which are separated by north–south and NNW–SSE-trending Variscan basement highs [18].

The Upper Rotliegend I of the SPB is commonly interpreted as a period that was mainly driven by compaction and thermal relaxation following a phase of intense early Rotliegend magmatism [15]. In contrast, the Upper Rotliegend II was marked by extensional faulting (Altmark I-IV tectonic phases: see Fig. 2.2; [19]). These tectonic phases may have constituted successive rifting events that preceded the opening of the Tethys and the Arctic-North Atlantic Oceans [20, 21].

In order to identify Rotliegend I and II depocentres suitable for hydrocarbon exploration, large-scale synsedimentary graben systems associated with regional transtensional deformation during the Late Carboniferous and Rotliegend were evaluated by Gast [22] for the Lower Saxony Basin; by Baltrusch and Klarner [23] for the NE German Basin and by Helmuth and Schretzenmayr [24] and Paul [25] for the Ilfeld Basin. Based on the interpretation of 3D seismic data from the Lower Saxony Basin, ~50 km SE of the study area, Gast and Gundlach [26] described two phases of graben formation: the early Rotliegend I graben formation was confined to WNW-trending strike-slip fault systems, as part of the early tectonic evolution of the SPB. The later Rotliegend II graben formation was associated with dominant W–E extension and minor strike-slip deformation along reactivated N-S-trending fault systems.

Using sequence stratigraphic methods, Börmann et al. [18] showed that the formation of NW–SE-striking pull-apart basins in the German North Sea sectors occurred contemporaneously to early Rotliegend volcanism, whereas N–S-striking normal faults were active during the late Rotliegend. Synsedimentary tectonic activity during the Upper Rotliegend II has also been documented by George and Berry [27] using data from 258 wells in the UK and Dutch sectors of the North Sea. They concluded that both strike-slip and dip-slip faulting occurred in the SPB during the late Rotliegend. In this work, the reactivation of NW–SE and NE–SW-trending pre-Permian basement lineaments within a W–E extensional regime controlled basin formation bounded by right-lateral and left-lateral oblique-slip faults.

In northern central Germany and the German North Sea sector, the Altmark I-IV tectonic pulses defined by Bachmann and Hoffmann [19] approximately correlate to a series of stratigraphic units that divide the Upper Rotliegend II into four upward-fining successions (Fig. 2.2; [28–31]): the Parchim, Mirow, Dethlingen and Hannover Formations, of which only the upper Dethlingen and Hannover Formations form tight gas reservoir units. Since the deposition of the Dethlingen Formation (Fig. 2.2), a perennial saline lake occupied the central part of the SPB [32]. Short-term but widespread marine incursions into the lake are represented by the Garlstorf and Schmarbeck Members of the Dethlingen Formation and by the Niendorf and Munster Members of the Hannover Formation [33, 34]. These marine incursions required a basinward topographic gradient, which resulted from subsidence of the basin floor below sea level, but have also been interpreted to have been caused by sea level highstands [35, 36]. Apart from lacustrine and rare marine deposits, fluvio-aeolian deposits form an important part of the sedimentary record of the Dethlingen and Hannover Formations in the study area: Gast [22] and Rieke et al. [37] concluded that aeolian sediment was supplied by the prevailing easterly winds, and the main fluvial sediment source areas were located in the Variscan hinterland in the south [11, 16].

Reservoir rocks of the study area (Figs. 1.2 and 2.2) are of fluvio-aeolian origin, including dune and wet to dry interdune deposits of the Wustrow and Bahnsen Members (Hannover Formation, Elbe Subgroup, Upper Rotliegend II). Aeolian dune deposits are (very) well bedded, fine to coarse grained, mature sandstones, which are characterized by semi-tabular cross-stratifications with foresets that show bimodal grain size lamination. Aeolian dune bases consist of sandstones with dip angles of 5°–15°, aeolian dunes show dip angles of 15°–35° at an average of ~22° [38]. Interdune units exhibit a variety of facies types that reflect the nature of the depositional surface at the time of accumulation [39, 40]. Dry sandflat deposits are characterized by mature, (very) well bedded to homogeneous, fine to coarse grained sandstones, which show bimodal grain size distributions and cross stratification with dip angles of less than 5°. They are dominated by low-angle translatent wind-ripple strata [39]. Wet sandflat deposits are very fine to fine grained, poorly sorted sandstones and siltstones with clay contents of 20–50 % [41]. Damp sandflat deposits are fine grained to medium grained sandstones with clay contents of less than 20 % [41]. Both are characterized by the occurrence of discontinuous, irregular to wavy argillaceous adhesion structures (e.g. ripples), small scale contortions (<0.2 m amplitude) and are often accompanied by irregular, lensoid to aligned concentrations of sand [38, 39].

Source rocks are Westphalian coals and the top seal is provided by Zechstein evaporites [42].

4.2 Data and Interpretation Methodology

The seismic interpretations presented are based on a pre-stack depth-migrated 3D seismic reflection volume and a (less-filtered) post-stack time-migrated seismic volume with a maximum N–S extent of 23.5 km and a maximum W–E extent of 17.5 km in the central part, covering an area of 293 km^2. The seismic reflection survey under study resulted from a merge of three data volumes, acquired in 1995, 1996 and 2001. After data merging, a trace length of 5 s and a sample interval of 4 ms were realized. Bin size was 25 $\times$ 25 m. Acquisition used Vibroseis, dynamite and airgun sources. The polarity is SEG normal. In addition, a regional 2D seismic line (post-stack time migration) with a W–E extent of $\sim$205 km, running through the southern part of the 3D volume (Fig. 4.1), was also analyzed.

Data from 14 wells, including digital wireline logs from seven wells, core data from four of these wells, Formation Micro Imaging/Formation Micro Scanning (FMI/FMS) logs from one of these wells and well reports and logs on paper from the other seven wells were also considered in this study. Six wells are located on a structural high in the eastern part of the survey area (Fig. 4.1).

Seismic interpretation took place in time and depth domains. For the structural interpretation, the depth domain data were preferred. Depth conversion was applied to the regional 2D line to enable it to be included into the interpretation. In total, 23 horizons, including seven horizons within the Permian stratigraphic interval, were interpreted over the 3D depth-migrated data set. Permian marker horizons are from top to base (see also Figs. 4.1, 4.2): Top Zechstein (positive amplitude = "+"); Top A2 Zechstein (negative amplitude = "−"); Top Z1 Zechstein (high amplitude, "+"); Top Rotliegend (high amplitude, "−"); Top Bahnsen (amplitude variation, "+") as top of the gas reservoir; Top Ebstorf ("+") as the base of the gas reservoir; and base of the Upper Rotliegend II ("+").

A total of 93 NNW–SSE and N–S-trending normal faults and 30 W-E-trending faults, controlling the Upper Rotliegend II interval, were interpreted on the basis of the 3D data. 18 additional Rotliegend faults were observed on the regional 2D line. Four salt domes and pillows were interpreted, one of which detailed on the basis of the 3D data. Isopach maps were generated for particular stratigraphic units bounded by seismic reflectors within the reservoir interval. 2D restoration and decompaction modelling was applied to the regional 2D seismic line and to separate 2D cross sections of the 3D seismic data. In addition, core analyses with an emphasis on sub-seismic scale structures and facies analyses were carried out.

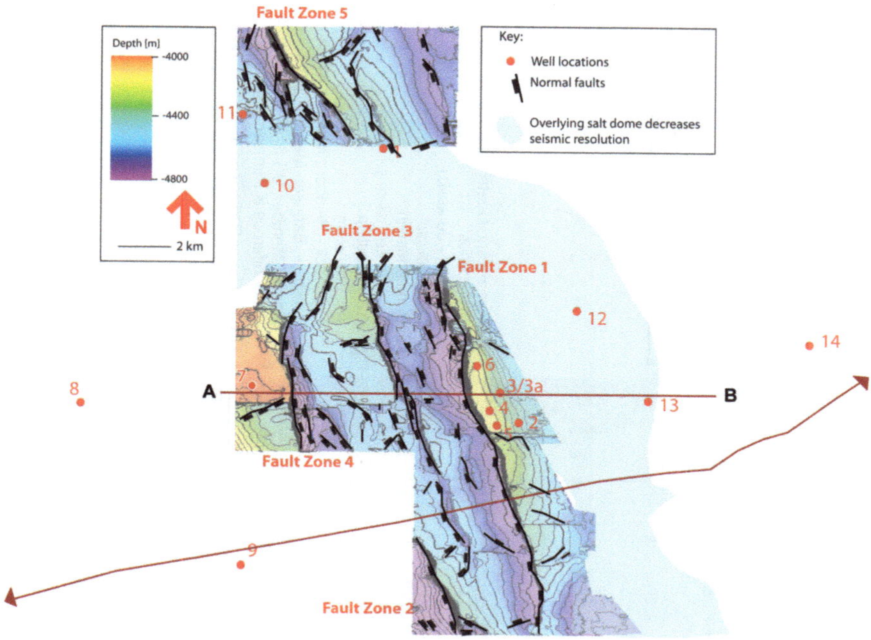

Fig. 4.1 Field development and structural features of the study area illustrated on the Top Rotliegend depth map. The W–E-trending *red line* depicts detailed interpreted seismic section shown in Fig. 4.2. Line with *arrowheads* displays orientation of regional 2D seismic line

4.3 Structural Subdivision of the Study Area

In the study area, a series of five NNW–SSE and N–S-striking normal fault zones with cumulative vertical offsets of up to 900 m have been identified, well illustrated by the Top Rotliegend depth map pattern (Fig. 4.1) and the structural analysis of seismic data (Fig. 4.3). These fault zones occur within a NNW–SSE to N–S-trending basin with subordinate NW–SE oriented half-grabens, internally marked by complex structures, such as relay ramps and en échelon step-fault zones with intervening ramp structures. The eastern footwall of the main graben zone (Fig. 4.1) is currently the sole gas-producing structure in the study area, and six wells here penetrate to the Rotliegend. From E to W, the following fault zones (Figs. 4.1, 4.3) were identified and interpreted:

Fault Zone 1 is composed of two major NNW–SSE-striking, down-to-the west normal fault segments (azimuth/dip: 225–330°/60–90°) with vertical offsets of up to 905 m. The faults are connected by a left-stepping overstep with releasing bends indicating a left-lateral transtensional stress system. The greatest vertical offsets developed in a fault-controlled sub-basin, bounded by a minor down-to-the east normal fault (Fig. 4.4). This is most likely a pull-apart structure caused by left-lateral transtension between the two major faults (Fig. 4.4).

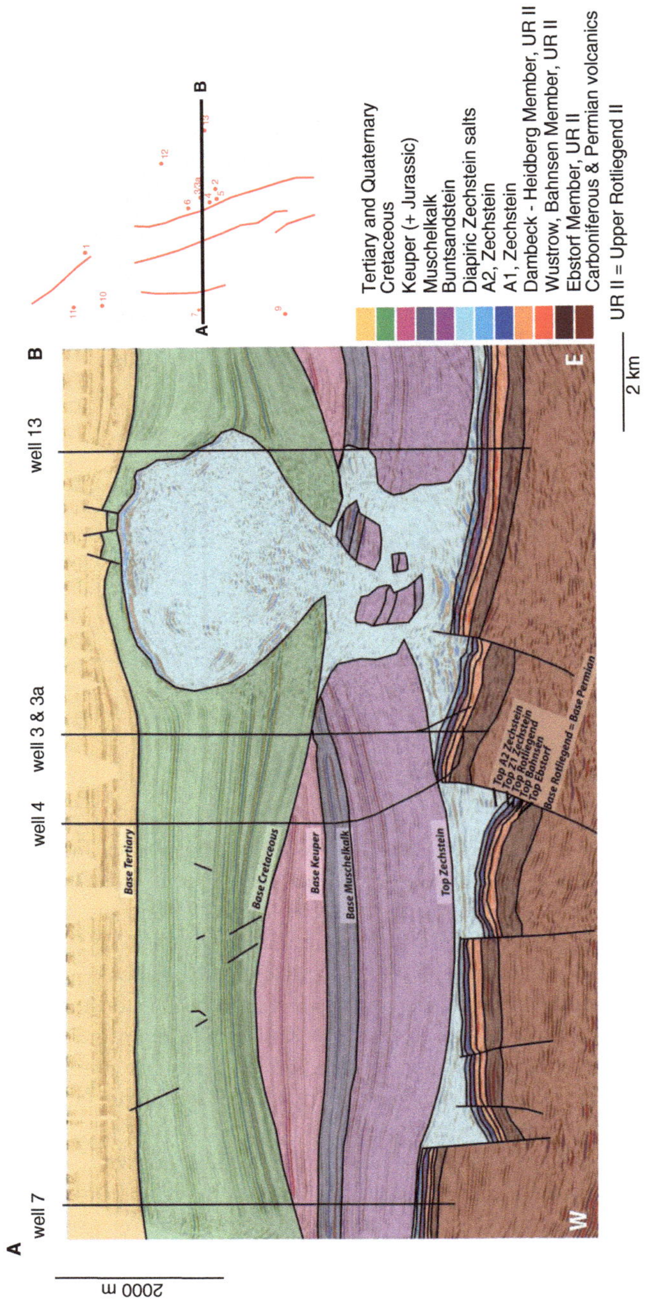

Fig. 4.2 W–E seismic section showing interpretation of main stratigraphic intervals and detailed Permian stratigraphy. See Fig. 4.3 for location of section

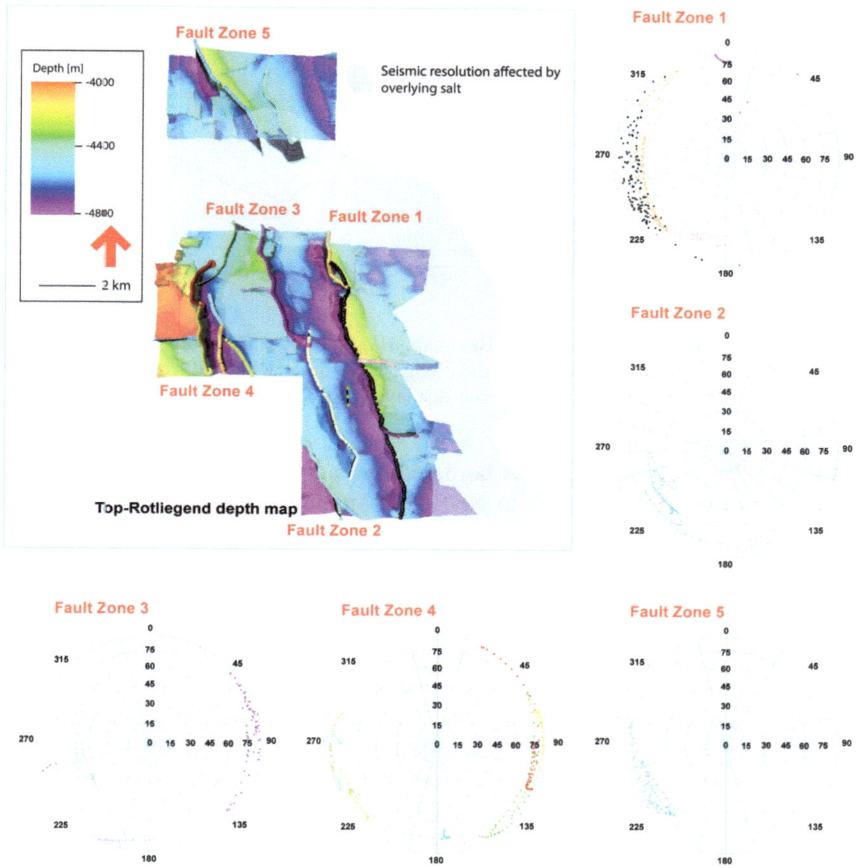

Fig. 4.3 *Upper left* Major fault zones of the study area (see text for explanation.). *Right* and *below* Orientations of fault planes constructed from structural analysis on seismic data plotted in Schmidt equal-area, upper-hemisphere contour diagrams

The NNW–SSE-striking Fault Zone 2 (Fig. 4.1) is a down-to-the-west normal fault, which vertically offsets the Top Rotliegend horizon by up to 575 m (azimuth/dip: 225–285°/60–80°).

Fault Zone 3 (Fig. 4.1) comprises a total of five NNW–SSE-trending fault segments with opposing dip directions in the centre of the graben system (Figs. 4.1, 4.3). In the south, three fault segments dip towards the W (azimuth/dip: 225–285°/45–90°), whereas the two northern fault segments dip towards the E (azimuth/dip: 45–135°/75–90°). The northern part of the fault shows vertical offsets of up to 300 m, whereas its southern part has maximum offsets of only 150 m. Because of the obvious change in dip directions along strike, the activity of this fault zone most probably comprised a considerable strike-slip component, causing oblique-slip movements.

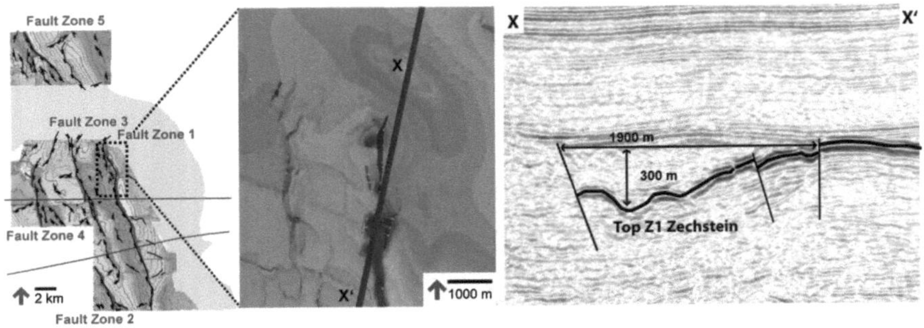

Fig. 4.4 Fault-controlled sub-basin. *Left* Basemap. *Middle* Depth map of Zechstein Z1 surface. *Right* Seismic section running parallel to the elongate sub-basin with faults depicted in thinner stroke and Top Z1-Zechstein seismic horizon displayed in thicker stroke

The western graben shoulder is bordered by Fault Zone 4 (Fig. 4.1), which is composed of three roughly N–S to NNE–SSW-trending, easterly dipping normal fault segments (azimuth/dip: 30–150°/80–90°) and associated westerly dipping antithetics. Vertical offsets of up to 650 m are developed in the area, where the southern part of Fault Zone 4 converges towards NW–SE-trending, down-to-the-west fault segments (azimuth/dip: 240–285°/80–90°) in the continuation of Fault Zone 2, forming the deepest part of a wedge-shaped narrow graben. The middle of the N–S to NNW–SSE-trending fault segments causes offsets of up to 450 m. It defines a graben shoulder where Top Rotliegend is at 4050 m depth, and the adjacent shallow basin is at a depth of ~4,500 m. The two southern, N–S-striking faults are connected by a left-stepping overstep, consistent with an inferred left-lateral transtensional stress regime. The footwall of Fault Zone 4 is not at a uniform depth, but segmented by a WSW–ENE-trending fault, achieving up to 500 m offset.

North of the salt dome (Fig. 4.1) Fault Zone 5 is located in an area where the northern continuations of Fault Zones 1 and 3 join. A major element of Fault Zone 5 is a NW–SE-trending, westward dipping normal fault (azimuth/dip: 190–270°/45–85°), which achieves vertical offsets of up to 400 m. The hangingwall of Fault Zone 4 comprises sets of smaller, roughly NW–SE-trending faults with variable dip directions.

4.4 Palaeo-Relief Analysis From Seismic Data

Although the Top Rotliegend depth map (Fig. 4.1) provides a reliable overview of the study area‘s present-day structural framework, it needs elimination of post-Permian tectonic effects to extract the fault-generated palaeo-topography, which existed at the time of Upper Rotliegend II to Zechstein deposition. In order to achieve this in the study area, differential thicknesses were calculated on the basis of Upper Rotliegend II and Zechstein isopach maps (Fig. 4.5) as well as of individual Upper Rotliegend subunits (Fig. 4.6).

In an extensional regime dominated by normal faulting, enhanced sediment thicknesses correspond to palaeo-hangingwall locations, whereas reduced thicknesses correspond to palaeo-footwalls (e.g. [43]. In lithologically uniform strata,

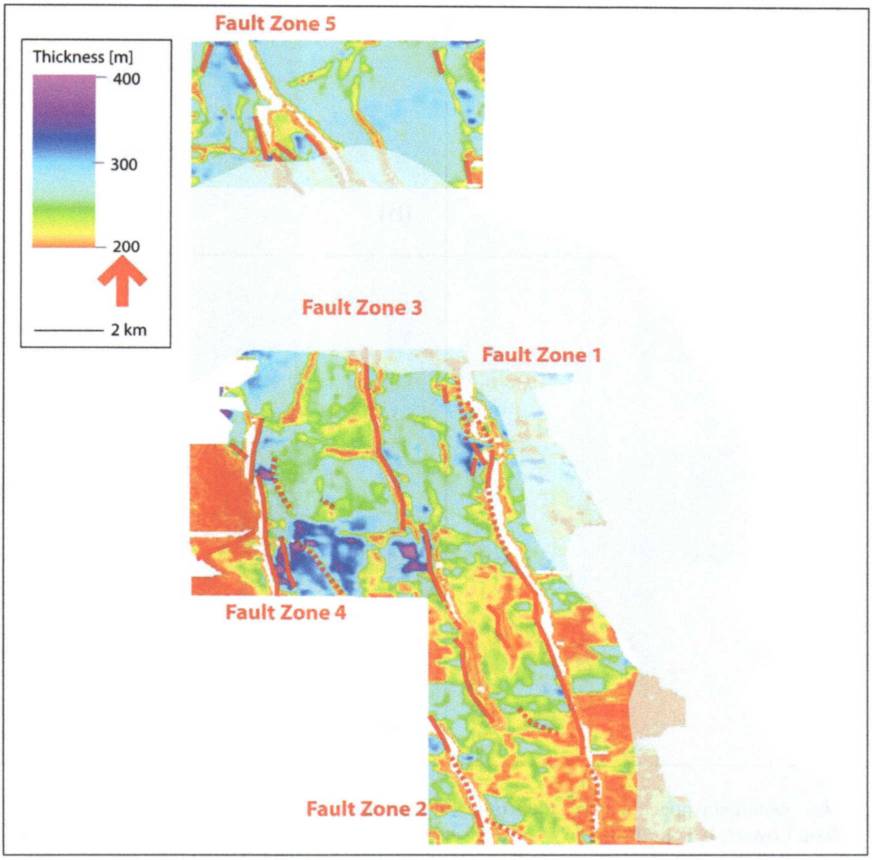

Fig. 4.5 Upper Rotliegend II isopach map; note that Rotliegend in the study area comprises Base Ebstorf until Top Heidberg. Faults with a paleorelief observed for the Rotliegend are marked in *red*. Overlying diapiring Zechstein salt, which affects the seismic resolution, is depicted in *transparent blue*. A *left*-lateral pull-apart sub-basin provides a localized depocentre associated with Fault Zone 1. The lowest sediment thickness developed on the footwall of Fault Zone 4, whereas the area of highest sediment thickness is located on the hangingwall of Fault Zone 4, forming a Rotliegend depocentre. Fault Zone 3 shows pronounced sediment thickness variations between foot and hangingwall in the *middle* part with high thicknesses of up to 400 m developed on the hangingwall

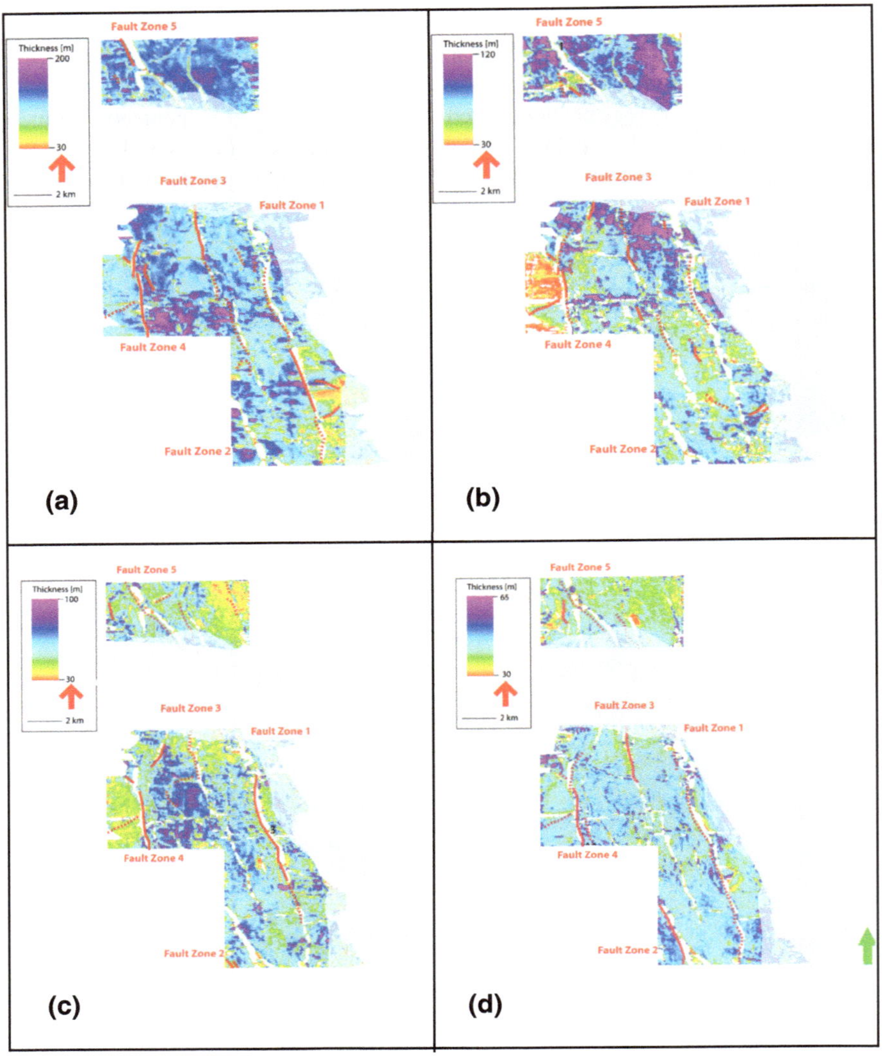

Fig. 4.6 Detailed isopach maps of the Rotliegend and lower Zechstein. **a** Base Upper Rotliegend II to Top Ebstorf. **b** Top Ebstorf to Top Bahnsen = reservoir isopach map. **c** Top Bahnsen to Top Rotliegend: Fault Zone 4 with paleorelief. **d** Zechstein Z1 isopach map. Enlarged sediment thicknesses can be observed for the area around the wedge-like structure between fault zones 1 and 3. It is displayed seperately in Fig. 4.9. Faults with a palaeo-relief observed are marked in *red*. Overlying Zechstein salt which affects the seismic resolution is depicted in *transparent blue*

even non-decompacted differential sediment thicknesses can therefore be used as proxy data for palaeo-relief reconstructions. The values presented here can be regarded as minimum values, because a consideration of decompaction would enhance fault-induced thicknesses variations.

Isopach pattern of Upper Rotliegend II strata (Fig. 4.5) do not allow an immediate distinction of palaeo-footwall and hangingwall positions on first sight. Instead, isopachs show only slightly enhanced sediment thicknesses on the hangingwalls of Fault Zone 1, of the northern segment of Fault Zone 2, of the northern and southern segments of Fault Zone 3 and of the Fault Zones 4 and 5. Thickness maxima are localized along the fault zones and occur particularly at stepovers, bends and wedges of faults, at releasing locations. Associated with Fault Zone 1, for instance, a localized pull-apart depocentre developed at a left stepover between two fault segments of the same trend. At Fault Zone 3, the Upper Rotliegend II reveals footwall thicknesses of 250–300 m, but a localized thickness maximum of ~400 m is achieved at a location of left stepover in the adjacent hangingwall. Along Fault Zone 4, an average thickness of 400 m occurs in the hangingwall, whereas the equivalent stratigraphic interval measures only 200 m thick in the footwall. Localized higher hangingwall thicknesses of up to 450 m occur at a left stepover of the north–south-trending fault segments and in the area where Fault Zones 2 and 4 join, suggesting an overall anastomosing fault pattern.

Detailed isopach maps of the intervals "Base Upper Rotliegend II to Top Ebstorf", "Top Ebstorf to Top Bahnsen" (reservoir interval), "Top Bahnsen to Top Rotliegend" and "Top Rotliegend to Top Z1 Zechstein" relate the structural framework of the study area to individual stratigraphic members and thus constrain the palaeo-relief evolution through time (Fig. 4.6a–d). During the Base Upper Rotliegend II to Top Ebstorf interval, Fault Zone 1 initially records a potential palaeo-relief of approximately 80 m in its southern half (Fig. 4.6a). Through time, fault activity in this zone shifted northwards (Fig. 4.6b) to develop differential hangingwall thicknesses of >50 m during the Top Bahnsen to Top Rotliegend interval (Fig. 4.6c) before levelling out during the early Zechstein (Fig. 4.6d). In the area of left stepover, fault-bounded hangingwall traps become particularly obvious.

In contrast, Fault Zone 2, at the SW edge of the survey, does not show any thickness variations during the studied Rotliegend intervals (Fig. 4.6a–c). Post-Rotliegend however, during the Z1 Zechstein, Fault Zone 2 developed a possible palaeo-relief of a few tens of metres (Fig. 4.6d).

Fault Zone 3 shows a maximum thickness contrast of >100 m between footwall and fault-bounded hangingwall traps in its central part, during the Base Upper Rotliegend II to Top Ebstorf interval (Fig. 4.6a). This zone of maximum differential thickness shifted northwards during deposition of the Top Ebstorf to Top Bahnsen interval (Fig. 4.6b). The fault zone was inactive during the remaining Rotliegend (Fig. 4.6c), but developed contrasting thicknesses during the early Zechstein (Fig. 4.6d).

Fault Zone 4 shows the largest contrasts of across-fault thicknesses throughout all investigated intervals. At the base of the studied succession, Fault Zone 4 records >100 m thickness difference between footwall and hangingwall (Fig. 4.6a, b). During Top Bahnsen to Top Rotliegend, the main depocentre moved eastwards towards Fault Zone 3 (Fig. 4.6c), and Fault Zone 4 finally lost its importance during the early Zechstein (Fig. 4.6d).

Fault Zone 5 solely shows only minor thickness highs on the hangingwall during Base Upper Rotliegend II to Top Ebstorf (Fig. 4.6a) and Top Bahnsen to Top Rotliegend (Fig. 4.6c); during deposition of Base Upper Rotliegend II to Top Ebstorf, enhanced thicknesses developed in an area where northern continuations of Fault Zones 1 and 3 merge (Fig. 4.6b).

4.5 Palaeo-Relief Analysis From Core and Log Data

From core data and well logs, the relative ages and facies of sedimentary units in the study area can be established. Braided stream dominated alluvium (Ebstorf Member: Fig. 2.2) and overlying fluvio-aeolian Wustrow and Bahnsen strata rest with pronounced unconformity on Early Rotliegend volcanic rocks or even Upper Carboniferous rocks, exposed at structural highs. Thus only the upper part of the Upper Rotliegend II stratigraphy is present in the study area at the southern margin of the SPB.

The unconformity between Top Carboniferous/Top Rotliegend volcanics and the overlying Ebstorf Member forms an obvious marker. Because Ebstorf strata represent the oldest Upper Rotliegend sediments in the study area, this unconformity defines the "Base Upper Rotliegend II" seismic marker.

Local areas of circular-shaped sediment thickness reductions are most obvious on the Base Upper Rotliegend II to Top Ebstorf isopach map (Fig. 4.6a), in the southern part of the Fault Zone 1 footwall. Thickness reductions at these locations may indicate remnant relief of the basaltic to andesitic volcanic rocks, which rest directly below the reservoir rock interval. Volcanic rocks at this level have only been identified in cores (Fig. 4.7), because their thicknesses (<5 m) are below seismic resolution at an average depth of 4,000 m. Nevertheless they show that volcanic activity occurred during (pre-)Ebstorf times.

Sets of aeolian strata of dune origin encountered by Wells 1–3a form the principal reservoir rocks and are assigned to the Upper Rotliegend II Wustrow and Bahnsen Members on the basis of seismic and well log interpretation. Individual dune sets, identified in the cored section, show maximum thicknesses of 3 m (Fig. 4.8). In most instances, these are encased in wet to dry interdune deposits. Limited dune set thicknesses together with pervasive cross-bedding and the considerable spread in palaeo-transport indicators suggest that the sand bodies most probably represent barchanoid dune forms. Figure 4.8 (lower left photograph) shows one of these barchanoid dune deposits and illustrates the variability in dip directions of foresets. Similar barchanoid dune sets have been identified by George and Berry [38] in the Upper Rotliegend of the southern North Sea (northern SPB). Prevailing barchanoid dune forms suggest limited sediment supply and/or restricted accommodation, which prevented the evolution of larger, more stable dune forms, such as transverse dunes [44]; or that the aeolian system built above preservation space and had little potential for being incorporated into the rock record [45].

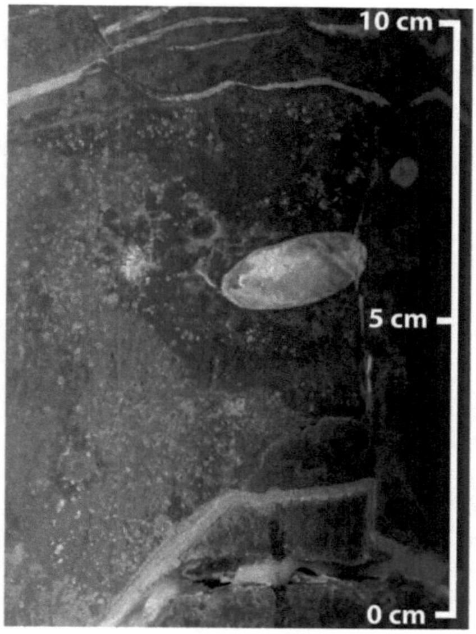

Fig. 4.7 Andesitic volcanics in core material of well 3, underlying the reservoir succession

4.6 Palaeo-Relief Uncertainty During Zechstein

Figure 4.9 shows arcuate reflectors interpreted as a salt removal structure in the Zechstein Z1 variance horizon slice and in seismic sections X-X' and Y-Y' across the Fault Zone 1 hangingwall. The structure is 2–2.5 km wide parallel to the fault zone (Fig. 4.9, section Y-Y'), and up to 880 m long perpendicular to the fault trace (Fig. 4.9, section X-X'), with a thickness of up to 350 m. Dip angles of 16° were measured. In addition, two smaller salt removal structures were identified adjacent to Fault Zones 2 and 5. The northern wedge, located at the hangingwall of Fault System 5, measures 1.4 km × 0.31 km and has a thickness of 110 m. The average stratal dip is 19°. The southern wedge is located on the hangingwall of Fault Zone 2 and has a fault-parallel extent of 0.7 km × 0.96 km, a thickness of 290 m, and an average dip of 16°.

Alternatively, these wedge-shaped features could be interpreted as footwall-sourced alluvial fans, which would suggest that considerable fault-generated relief was present during deposition of Zechstein strata. There is no well control yet which would exclude one of the presented interpretations. Assuming that the wedge-shaped feature along Fault Zone 1 represents indeed an alluvial fan, this would imply a palaeo-relief of ~220 m developed during the Zechstein Z2 to Z3 interval. The cumulative thickness of the wedge-shaped feature is ~350 m; a minimum of ~220 m of Zechstein fault offset along the hangingwall of Fault Zone 1 was calculated although only a palaeo-relief of ~80 m can be detected for the Upper Rotliegend II. Because during the Upper Rotliegend II no palaeo-relief

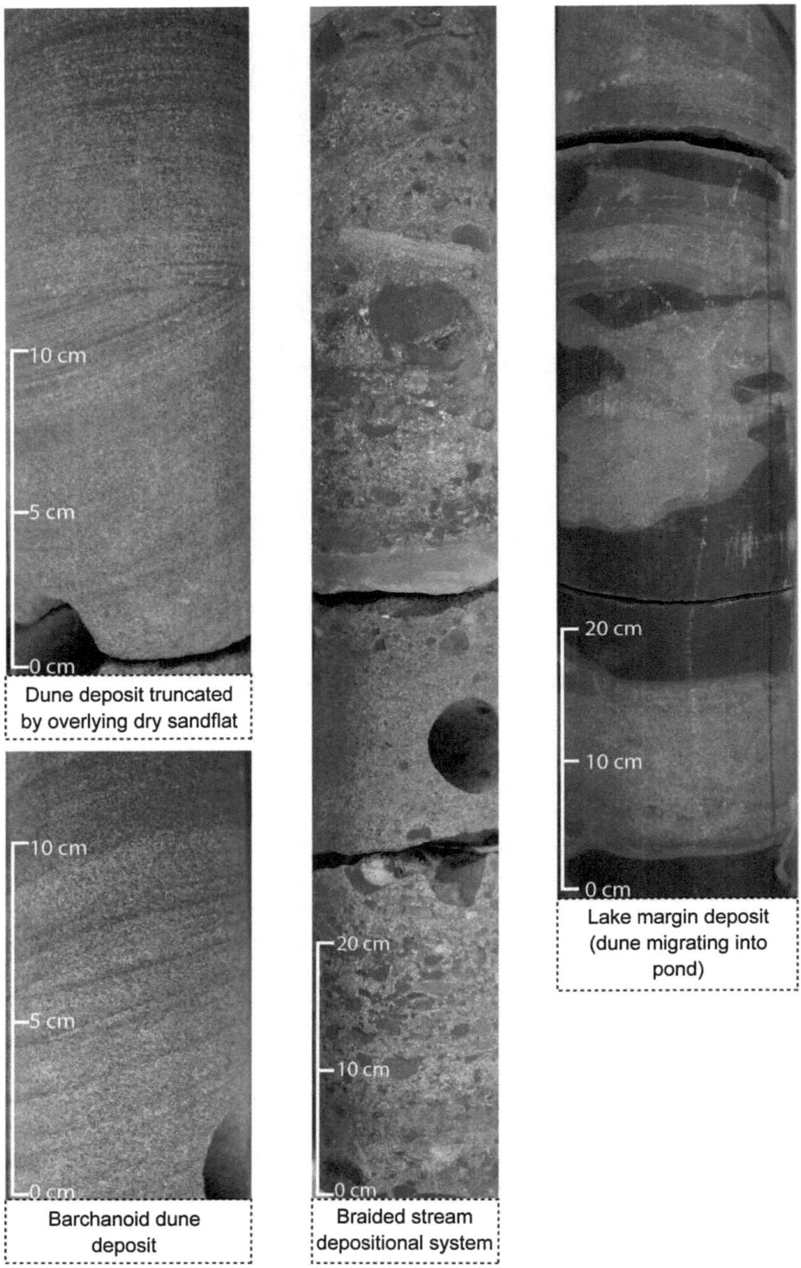

Fig. 4.8 Core material of well 3. The core succession is dominated by wet to dry aeolian deposits. Volcanics and their overlying breccias and the deposits of braided stream dominated alluvial fans are situated in the *lower* part of the core beneath the reservoir rocks

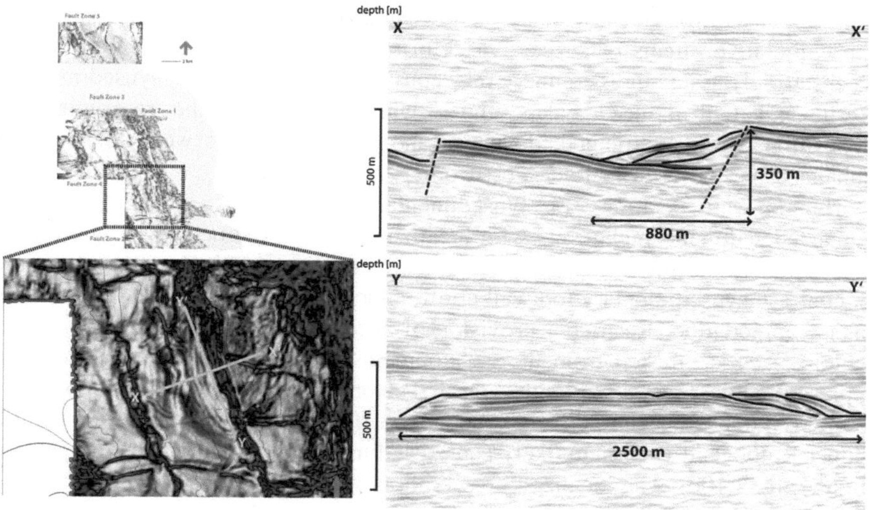

Fig. 4.9 Arcuate salt removal wedge on Top Z1-Zechstein variance horizon slice and in seismic sections. *Upper left* Top Z1-Zechstein variance horizon slice with overlying salt and fault zone annotations. *Lower left* Z1 Zechstein variance horizon slice zoomed on the outlines of the wedge-like structure. X-X′ and Y-Y′ location of seismic sections displayed on the *right*. *Upper right* Seismic section perpendicular to Fault Zone 1 that borders the wedge-like structure. Different salt removal cycles or reflectors of floaters can be observed. *Lower right* Seismic section cut parallel to Fault Zone 1. The reflectors outline a lobate structure with removal cycles or floater reflectors to the N

is identified in association with Fault Zone 2 and only minor palaeo-relief is observed along Fault Zone 5, an interpretation of the wedges as alluvial fans would imply that ~290 m of fault-induced topography along Fault Zone 2 (and ~110 m along Fault Zone 5), was established during earliest Zechstein deposition.

The most likely alternative is that the wedge-shaped features may have been initiated by salt and carbonate withdrawal, causing the formation of voluminous salt removal breccia. Reflector geometries would then suggest at least two gravity-induced mass transport events or, more likely, salt removal cycles (Fig. 4.9). Warren [46] provides a detailed description of salt dissolution processes and/or withdrawal, leading to the formation of residual breccia. Goudie [47] proposed that the removal of salt may leave amphitheatrical hollows, containing abundant collapsed material and exotic blocks.

4.7 Discussion

One key observation of the structural analysis provided is, that the total offsets of the fault zones interpreted on seismic data (Fig. 4.1) are much higher compared to the observed palaeo-relief of the Rotliegend interpreted from the isopach data

(Figs. 4.5 and 4.6). This is because the present-day structural configuration records the cumulative effects of successive deformation phases during the Rotliegend, Zechstein, Triassic, Jurassic and Cretaceous (e.g. [18, 20]). The synsedimentary Rotliegend fault-induced palaeo-highs, which influenced the reservoir rock deposition, were re-arranged, displaced and the fault offsets were considerably enhanced during multi-tectonic post-Rotliegend overprint.

Hangingwall positions of synsedimentary fault zones are registered by enhanced sediment thicknesses with localized pull-apart structures providing the most efficient sediment traps, favouring sediment accumulation and preservation. In contrast, synsedimentary footwall uplift reduces or even precludes the provision of accommodation. Restricted provision of accommodation, respectively preservation space is inferred from the development of dunes on a local palaeo-high, representing the footwall of Fault Zone 1. Tectonically induced palaeo-topography is inferred to have been relatively subtle there. High sedimentation rates in the arid Upper Rotliegend II climate likely caused sediments to cover the existing palaeo-highs (Fig. 4.10b). The localized preservation of dune deposits in the footwall of Fault Zone 1 may as well relate to the formation of sub-seismic scale footwall collapse compartments. Other factors, which might have supported the accumulation and subsequent preservation of dunes in such an atypical setting may include sand stabilization by increased moisture content or formation of surface crusts.

In the context of the predominantly easterly to north-easterly wind directions, inferred from publications by Gast [22] and Rieke et al. [37] and from the analyzed FMI/FMS logs, the palaeo-relief associated with Fault Zone 1 may have provided a leeside trap. Thus, thicker aeolian sand accumulations, evening the existing palaeo-relief, can be expected to occur in hangingwall positions west of Fault Zone 1. A limiting factor on the growth of the Rotliegend aeolian dune forms may have been high wind velocities on the up-wind side of elevated palaeo-topography (e.g. the footwall of Fault Zone 1), causing sediment bypass or deflation there [38]. A well known modern example for the development of extensive deflation surfaces is the coastal Namib Desert, Namibia, which is dominated by strong onshore winds [48, 49]. In contrast, down-wind of the footwall-sourced fans, enhanced dune growth was probably favoured through the provision of local sediment sources.

A remarkable palaeo-elevation of at least $\sim$150 m is identified in the central part of Fault Zone 3 (Figs. 4.5, 4.6), which has been considered to include a considerable strike-slip component. On the basis of the isopach maps, the existence of palaeo-relief can be postulated for the entire Upper Rotliegend II (Figs. 4.5, 4.6), and the fault systems are therefore inferred to have been active during the entire period. Upper Rotliegend II thickness maxima are clearly fault-defined and relate to a release location in an area of left overstep between southern and northern segments of Fault Zone 3.

Maximum provision of accommodation, due to the existence of up to 250 m of fault-induced palaeo-relief, can be observed at the hangingwall of the eastward-dipping Fault Zone 4 in the western part of the studied area (Figs. 4.5, 4.6). Favoured by this palaeo-relief, an Upper Rotliegend II succession up to 250 m thick was deposited on the hangingwall, located up-wind of the palaeo-high. Log

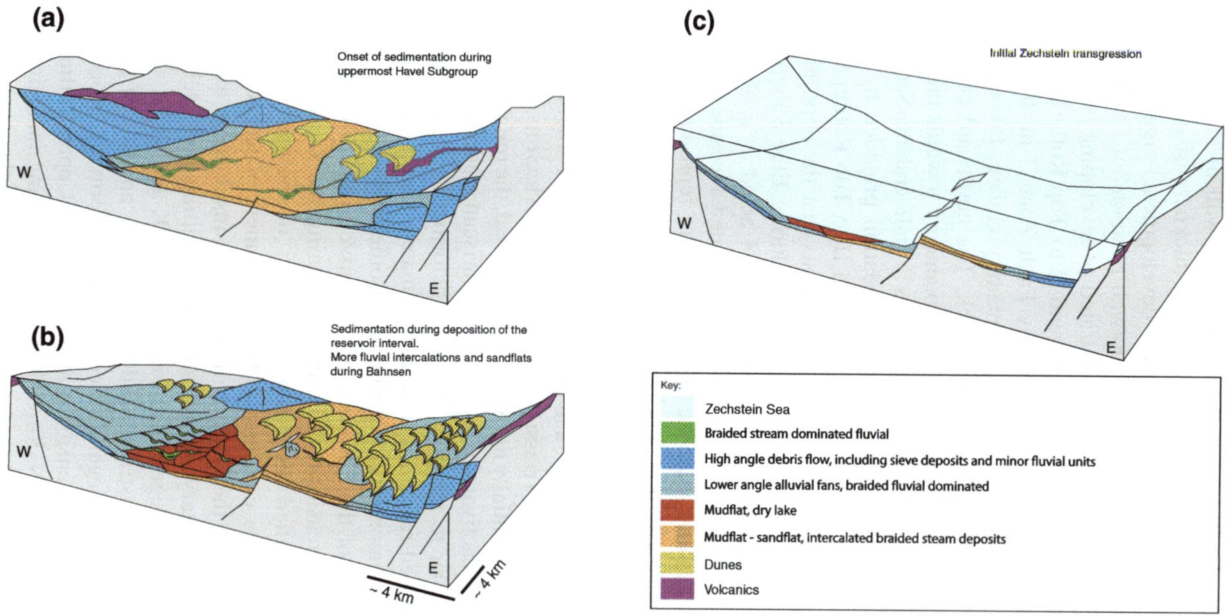

Fig. 4.10 Schematic model of the sedimentary development of the study area. **a** From Base Upper Rotliegend II to Top Ebstorf volcanics were erupted locally and alluvial fans were shedded from the Carboniferous graben and halfgraben shoulders. **b** During the reservoir interval (Top Ebstorf to Top Bahnsen) a perennial lake existed at the Rotliegend depocentre. On the shoreline of the perennial dry lake aeolian dunes developed. Higher dunes accumulated on the hangingwall of fault zones. They partly covered the paleorelief. Smaller dune successions developed on the footwalls. Palaeo-topography and potential fault activity was subtle. **c** Study area after the Zechstein transgression

interpretation of Well 7 (Fig. 4.1), located on the footwall of Fault Zone 4, indicates the presence of sandflats with a cumulative thickness of up to 65 m in the Wustrow and Bahnsen Members. Sandflat development suggests high wind velocities and limited accommodation in footwall positions of Fault Zone 4.

In Fig. 4.10, Fault Zones 3 and 4 are illustrated as developing prominent scarps during deposition of the Base Upper Rotliegend II to Top Rotliegend units. The resulting sedimentary depocentre, bounded by Fault Zone 3 in the E and Fault Zone 4 in the W, is interpreted as a synsedimentary graben, favouring accumulation of Rotliegend deposits up to 400 m thick. The northward continuation of Fault Zone 4 shows no evidence of a conjugate boundary towards the east and the associated depocentre is therefore interpreted as a half-graben. Sediments may continuously have accumulated in the accommodation provided by the graben/halfgraben area during the Base Upper Rotliegend II to Top Rotliegend interval. The interpretation of the isopach maps also suggests that the depocentre next to Fault Zone 3 was progressively shifted eastwards during Elbe Subgroup deposition. The palaeo-relief associated with Fault Zone 4 acted as a windward trap, whereas the positive relief along Fault Zone 3 represented a leeside trap for aeolian sediments. Within the study area, the latter depocentre represents the most pronounced topographic low during the Upper Rotliegend II, and was most likely affected by a fluctuating groundwater table or even intermittent periods of ephemeral flooding, eventually leading to localized deposition of muds.

The orientation of Fault Zone 4, and of the central part of Fault Zone 3, is nearly N–S and does not fit with the Triassic stress field under which most of the NNW-SSE-trending faults preferentially developed [18]. Instead, the roughly N–S oriented faults may relate to the reactivation of syn- to late-Variscan grabens which originally developed perpendicular to the main Variscan deformation front [18]. Gast and Gundlach [26] described reactivation of approximately north–south oriented tectonic elements as a common process during late Rotliegend graben formation.

Renewed movement of the European plate relative to the African plate led to the reactivation of transtensional fault systems during the Upper Rotliegend II [19]. Glennie [50] identified the combined effects of thermal subsidence and strike-slip faulting as major controls on the development of late Rotliegend subbasins in the Southern and the Northern Permian Basins. George and Berry [27] concluded that synsedimentary dextral strike-slip faulting and cyclical climatic changes controlled late Rotliegend deposition in the UK and Dutch sectors of the SPB. Based on frictional constraints and slip-tendency analysis of fault-induced palaeo-relief of the entire Rotliegend succession in NE Germany, Moeck et al. [51] suggested that the E-W transtensional syndepositional Upper Rotliegend stress regime was best expressed by N–S-trending normal faulting including strike-slip components. This transtensional regime persisted during the entire Upper Rotliegend II.

Palaeo-topography, the majority of which is related to N–S-trending normal faults, can be proven for all fault zones in the study area, except Fault Zone 2. All of these fault zones show evidence for syndepositional left-lateral oblique-slip movements, indicated by left-stepping oversteps, pull-apart structures and variable dip directions of strata within a single fault zone.

In contrast, thickness variations of up to 750 m in the late Zechstein cannot solely be attributed to synsedimentary tectonic activity, as contemporaneous salt movements have to be considered. Salt movements, starting during the Early Triassic [52] and raft tectonics, initiating during the late Triassic, played an important role and modified original thicknesses of Zechstein salt and carbonate accumulations. However, salt movement was probably initiated by sediment loading, with differential values controlled by tectonic activity. The "Tubantian I" tectonic event (Fig. 2.2; [53, 54]), which is well known in the Netherlands, may have played an important role in the initiation of salt movement. Tectonic events contemporaneous with the onset of Zechstein deposition are also known from the southern North Sea [34]. Ziegler [20] and Vejbæk [55] proposed that the Zechstein evaporites were deposited as a consequence of fault-controlled subsidence. The seismic interpretation results in this chapter suggest that minor fault-related relief was established during the earliest Z1 Zechstein, and deposition was tectonically unaffected during the later Zechstein. As a consequence, overprint by salt movements does not allow reliable reconstruction of palaeo-topography from thickness variations of late Zechstein deposits.

4.8 Conclusions

This study allows conclusions to be drawn regarding the synsedimentary fault-induced palaeo-relief that influenced sedimentary deposition in the study area during the Upper Rotliegend II.

1. The study area records evidence for left-lateral oblique-slip activity of N–S to NNW-SSE-trending faults contemporaneous with Upper Rotliegend II deposition. Aeolian dune accumulation and preservation potential is favoured in palaeo-hangingwall positions of normal faults or more localized pull-apart and overstep structures at releasing locations of strike-slip and oblique-slip faults. In contrast, palaeo-footwall positions are not a setting, where dunes would typically construct and is an even less likely setting for their accumulation and subsequent preservation. The exceptional and localized preservation of dune deposits in palaeo-footwall positions (Fault Zone 1) may relate to subtle palaeo-relief, the formation of sub-seismic scale footwall collapse compartments and increased moisture content at the sediment surface.
2. The Altmark IV tectonic event, which comprised normal faulting during deposition of the Upper Rotliegend II Ebstorf and Wustrow Members, is identified as trigger for the onset of deposition in the study area. Later, during the Upper Rotliegend II, more localized tectonic phases caused additional fault offsets. Latest Rotliegend activity on N–S-trending faults is proven for the area of East Frisia by this study.

3. Upper Rotliegend II and/or older faults were repeatedly reactivated under changing stress regimes, in particular during Triassic-Jurassic faulting, often causing the development of cumulative fault offsets and a post-depositional expansion of the original fault extent. In most instances, the mainly NW–SE-trending faults (e.g. Fault Zone 1 and its suspected northern continuation by Fault Zones 5 and 3) successively enlarged their extents and finally became connected. The present-day structural configuration records the cumulative effects of successive multi-tectonic overprint of deformation phases during the Rotliegend, Zechstein, mainly Triassic, Jurassic and Cretaceous.
4. N–S-trending faults (e.g. Fault Zone 4) and NNW–SSE-oriented fault segments (e.g. Fault Zones 1 and 3) show evidence of syndepositional fault activity and related fault-induced topography during the entire Upper Rotliegend II. In contrast, Fault Zone 2 does not show differential footwall and hangingwall sediment thicknesses during the Permian. Due to its NW–SE orientation and the westward dip, I assume that this fault zone developed entirely under a younger stress regime. As fault-induced palaeo-relief is mainly postulated for the N–S-trending faults and fault zone segments, I anticipate that thick sandstone reservoir rock deposits are preferentially located in hangingwall positions of these faults. NNW–SSE-trending faults in most instances originated from later connection of N–S-trending fault segments and can therefore be regarded as partly having fault-induced palaeo-relief developed during the deposition of the Upper Rotliegend II. According to the model, westward dipping NW–SE fault zones would then encounter less thick sandflat deposits.
5. The development of local depocentres, which provided accommodation for clastic reservoir rocks during the Upper Rotliegend II were associated with complex relay ramps (Fault Zone 3) and pull-apart sub-basins (Fault Zone 1), suggesting left-lateral transtensional stresses. Syndepositional extensional movements along N–S to NW–SE-trending fault zones and fault segments thus were not pure dip-slip but included a considerable oblique-slip component.
6. Rotliegend faults, that underwent a tectonic reactivation during the Zechstein, may have been a pre-requisite for the initiation of salt movement. Decoupling of late Permian Zechstein salts and clays was conductive to the already existing structural grain, and caused the structural style of the post-Zechstein units to vary significantly from that of the Rotliegend.
7. The analysis of fault-induced palaeo-relief through time represents a useful tool to unravel depositional environments that underwent multi-phase tectonic histories. Furthermore, it supports the extra- and interpolation between wells in undrilled areas and helps to point out suitable locations for hydrocarbon reservoir rocks.

References

1. Law BE, Curtis JB (2002) Introduction to unconventional petroleum systems. AAPG Bulletin 86(11):1851–1852
2. Holditch SA (2006) Tight gas sands. J Petrol Technol 58(6):86–93
3. Gaupp R, Matter A, Platt J, Ramsayer K, Walzebuck JP (1993) Diagenesis and fluid evolution in deeply buried Permian (Rotliegende) gas reservoirs. NW Germany, AAPG Bulletin 77(7):1111–1128
4. Gaupp R, Solms M (2005) Palaeo oil- and gasfields in the Rotliegend of the North German basin: effects upon hydrocarbon reservoir quality (Paläo-Öl- und Gasfelder im Rotliegenden des Norddeutschen Beckens: Wirkungen der KW-Migration auf die Speicherqualitäts-Entwicklung.). In: Gaupp R (ed) DGMK Research report 593: Tight gas reservoirs—natural gas for the future: DGMK Celle, p 242
5. Lehner FK, Pilaar WF (1991) On a mechanism of clay smear emplacement in synsedimentary normal faults. In: AAPG 1991 annual convention with DPA/EMD divisions and SEPM, AAPG Bulletin, 75(3), p 619
6. Lindsay NG, Murphy FC, Walsh JJ, Watterson J (1993) Outcrop studies of shale smears on fault surfaces. In: Flint SS, Bryant ID (eds) The geological modelling of hydrocarbon reservoirs and outcrop analogues. Special publication of the international association of sedimentologists, vol 5, p 113–123
7. Knipe RJ (1997) Juxtaposition and seal diagrams to help analyze fault seals in hydrocarbon reservoirs. AAPG Bulletin 81:187–195
8. Stollhofen H (1998) Facies architecture variations and seismogenic structures in the carboniferous-permian saar-nahe basin (SW Germany): evidence for extension-related transfer fault activity. Sed Geol 119:47–83
9. Drong HJ, Plein E, Sannemann D, Schuepbach MA, Zimdars J (1982) Der Schneverdinger-Sandstein des Rotliegenden; eine aeolische Sedimentfüllung alter Grabenstrukturen. Zeitschrift der Deutschen Geologischen Gesellschaft 133(3):699–725
10. Glennie KW (1986) Development of NW Europe‘s Southern Permian gas basin. In: Brooks J, Goff JC, van Horn B (eds) Habitat of paleozoic gas in N.W, Europe, Geological society of London, special publication, vol 23, p 3–22
11. Glennie KW (1990a) Rotliegend sediment distribution; a result of late carboniferous movements. In: Hardman RFP Brooks J (eds) Proceedings of tectonic events responsible for Britain‘s oil and gas reserves: geological society of London, Special publications, vol 55, p 127–138
12. Plein E (1995) Stratigraphie von Deutschland I, Courier Forschungsinstitut Senckenberg, v. 183
13. Strömbäck AC, Howell JA (2002) Predicting distribution of remobilized aeolian facies using sub-surface data: the Weissliegend of the UK Southern North Sea. Petrol Geosci 8:237–249
14. Legler B (2005) Faziesentwicklung im Südlichen Permbecken in Abhängigkeit von Tektonik, eustatischen Meeresspiegelschwankungen des Proto-Atlantiks und Klimavariabilität (Oberrotliegend, Nordwesteuropa): Schriftenreihe der Deutschen Gesellschaft für Geowissenschaften, vol 47, p 103
15. van Wees JD, Stephenson RA, Ziegler PA, Bayer U, McCann T, Dadlez R, Gaupp R, Narkievicz M, Bitzler F, Scheck M (2000) On the origin of the Southern Permian basin, central Europe. Mar Pet Geol 17:43–59
16. Plein E (1993) Bemerkungen zum Ablauf der palaeogeographischen Entwicklung im Stefan und Rotliegend des Norddeutschen Beckens. Observations on Stephanian and Rotliegendes palaeogeography in the North German Basin. In: Zur Geologie und Kohlenwasserstoff-Fuehrung des Perm im Ostteil der Norddeutschen Senke. Geology and hydrocarbon potential of the Permian rocks of the eastern North German Basin: Geologisches Jahrbuch. Reihe A: Allgemeine und Regionale Geologie BR Deutschland und Nachbargebiete, Tektonik, Stratigraphie, Palaeontologie, vol 131 p 99–116

17. McCann T (1998) The Rotliegend of the NE German basin: background and prospectivity. Petrol Geosci 4:17–27
18. Börmann C, Gast R, Görisch F (2006) Structural and sedimentological analysis of an early late Rotliegendes graben based on 3D seismic and well log data, German North Sea. Petrol Geosci 12:195–204
19. Bachmann GH, Hoffmann N (1997) Development of the Rotliegend basin in northern Germany. Geologisches Jahrbuch. Reihe D: Mineralogie, Petrographie, Geochemie, Lagerstaettenkunde 103:9–31
20. Ziegler PA (1990) Geological atlas of Western and central Europe. Shell, 2nd edn. The Hague, p 239
21. Stollhofen H, Bachmann NGH, Barnasch J, Bayer U, Beutler G, Franz M, Kästner M, Legler B, Mutterlose J, Radies D (2008) Upper Rotliegend to early cretaceous basin development. In: Littke R, Bayer U, Gajewski D, Nelskamp S (eds) Dynamics of complex intracontinental basins; the central European basin system. Springer, Berlin, pp 181–210
22. Gast RE (1988) Rifting im Rotliegenden Niedersachsens. Rifting in the Rotliegendes of Lower Saxony: Die Geowissenschaften Weinheim 6(4):115–122
23. Baltrusch S, Klarner S (1993) Rotliegend-graeben in NE-brandenburg. Graben formation in the Rotliegend of NE-brandenburg. Zeitschrift der Deutschen Geologischen Gesellschaft 144(1):173–186
24. Helmuth HJ, Schretzenmayr S (1995) Zur raum-zeitlichen Genese der Gräben.In:Plein E (ed) Stratigraphie von Deutschland I; Norddeutsches Rotliegendbecken – Rotliegend-Monographie Teil II. Courier Forschungsinstitut Senckenberg, vol 183, p 169–174
25. Paul J (1999) Evolution of a Permo-Carboniferous basin; the Ilfeld Basin and its relationship to adjoining Permo-Carboniferous structures in central Germany. Neues Jahrbuch fuer Geologie und Palaeontologie. Abhandlungen, vol 214(1–2), p 211–236
26. Gast R, Gundlach T (2006) Permian strike-slip and extensional tectonics in lower Saxony, Germany. Zeitschrift der Deutschen Gesellschaft für Geowissenschaften 157(1):41–55
27. George GT, Berry JK (1997) Permian (Upper Rotliegend) synsedimentary tectonics, basin development and palaeogeography of the southern North Sea. In: Ziegler K, Turner P, Daines SR (eds) Petroleum geology of the southern North Sea; future potential. Geological society of London, Special publications, vol 123, p 31–61
28. Hoffmann N, Kamps H-J, Schneider J (1989) Neuerkenntnisse zur Biostratigraphie und Paläodynamik des Perms in der Nordostdeutschen Senke – ein Diskussionsbeitrag. Z Angew Geol 35:198–207
29. Hoffmann N (1990) Zur paläodynamischen Entwicklung des Präzechsteins in der Nordostdeutschen Senke. Niedersächsische Akademie der Wissenschaften, Geowissenschaftliche Veröffentlichung 4:5–18
30. Gebhardt U, Schneider J, Hoffmann N (1991) Modelle zur Stratigraphie und Beckenentwicklung im Rotliegenden der Norddeutschen Senke. Geol Jahrb A127:405–427
31. Gast R (1995) Sequenzstratigraphie. In: Plein E (ed) Stratigraphie von Deutschland I; Norddeutsches Rotliegendbecken – Rotliegend-Monographie Teil II. Courier Forschungsinstitut Senckenberg, vol 183, p 47–54
32. Gast R, Gaupp R (1991) The sedimentary record of the Late Permian saline lake in N.W. Germany. In: Renaut, RW, Last WM (eds) Sedimentary and Paleolimnological Records of Saline Lakes. Natl Hydrol Res Inst, Saskatoon, SK, Canada, p 75–86
33. Gast R, Gebhardt U (1995) Elbe Subgruppe. In: Plein E (ed) Stratigraphie von Deutschland I; Norddeutsches Rotliegendbecken – Rotliegend-Monographie Teil II. Courier Forschungsinstitut Senckenberg, vol 183, p 121–145
34. Legler B, Gebhardt U, Schneider JW (2005) Late Permian non-marine – marine transitional profiles in the central Southern Permian Basin. Int J Earth Sci 94:851–862
35. Stemmerik L, Ineson JR, Mitchell JG (2000) Stratigraphy of the Rotliegend group in the Danish part of the Northern Permian basin, North Sea. J Geolo Soc 157:1127–1136

36. Stemmerik L (2001) Sequence stratigraphy of a low productivity carbonate platform succession: the Upper Permian Wegener Halvø Formation, Karstryggen Area, East Greenland. Sedimentology 48:79–97
37. Rieke H, Kossow D, McCann T, Krawczyk C (2001) Tectono-sedimentary evolution of the northernmost margin of the NE German basin between uppermost Carboniferous and Late Permian (Rotliegend). Geol J 36(1):19–38
38. George GT, Berry JK (1993) A new palaeogeographic and depositional model for the Upper Rotliegend of the UK sector of the Southern North Sea. In: North CP, Prosser DJ (eds) Characterization of fluvial and Aeolian reservoirs. Geological society of London, Special publication, vol 73, p 291–319
39. Mountney NP, Jagger A (2004) Stratigraphic evolution of an erg margin aeolian system: the permian cedar mesa sandstone, SE Utah, USA. Sedimentol 51:713–743
40. Mountney NP, Russell AJ (2009) Aeolian dune field development in a water table-controlled system: Skeiđarársandur, southern Iceland. Sedimentol 56:2107–2131
41. Amthor JE, Okkerman J (1998) Influence of early diagenesis on reservoir quality of Rotliegende sandstone, Northern Netherlands. AAPG Bulletin 82:2246–2265
42. Schwarzer D, Littke R (2007) Petroleum generation and migration in the 'Tight Gas' area of the Germany Rotliegend natural gas play: a basin modelling study. Petrol Geosci 13:37–62
43. Eisbacher GH (1996) Einführung in die Tektonik. 2. neu bearbeitete und erweiterte Auflage. Stuttgart: Enke, p 374
44. Mountney NP (2006) Eolian Facies Models, Facies models revisited. In: Posamentier H, Walker RG(ed), SEPM Mem, vol 84, p 19–83
45. Kocurek G, Havholm KG (1993) Eolian sequence stratigraphy; a conceptual framework. In: Siliciclastic sequence stratigraphy; recent developments and applications. AAPG Memoir, vol 58, p 393–409
46. Warren J (1999) Evaporites: their evolution and economics. Blackwell Science, Oxford, p 438
47. Goudie AS (1989) Salt tectonics and geomorphology. Prog Phys Geogr 13:597
48. Corbett I (1993) The modern and ancient pattern of sandflow through the southern Namib deflation basin. In: Aeolian sediments, ancient and modern. Pye K, Lancaster N (ed) Special publication of the international association of sedimentologists, vol 16, p 45–60
49. Krapf C, Stollhofen H, Stanistreet IG (2003) Contrasting styles of ephemeral river systems and their interaction with dunes of the Skeleton Coast erg (Namibia). Quatern Int 104:41–52
50. Glennie KW (1997) Recent advances in understanding the southern North Sea Basin, a summary. In: Ziegler K, Turner P, Daines SR (eds) Petroleum geology of the southern North Sea; future potential. Geological Society of London, Special publications, vol 123, p 17–29
51. Moeck I, Schandelmeier H, Holl H-G (1999) The stress regime in a Rotliegend reservoir of the northeast German Basin. Geologische Rundschau =. Int J Earth Sci 98(7):1643–1654
52. Mohr M, Kukla PA, Urai JL, Bresser G (2005) Multiphase salt tectonic evolution in NW Germany; seismic interpretation and retro-deformation. Int J Earth Sci 94(5–6):917–940
53. Geluk M (1999) Late Permian (Zechstein) rifting in the Netherlands; models and implications for petroleum geology. Petrol Geosci 5:189–199
54. Geluk MC (2000) Late Permian (Zechstein) carbonate-facies maps, the Netherlands. Geologie en Mijnbouw, Netherlands Journal of Geosciences, v. 79(1):17–27
55. Vejbæk OV (1990) The Horn Graben, and its relationship to the Oslo Graben and the Danish Basin in Rift zones in the continental crust of Europe; geophysical, geological and geochemical evidence; Oslo-Horn Graben. Neumann Tectonophysics 178(1):29–49

Chapter 5
The Panamint Valley, Western US: A Field Analogue for the Sedimentary Facies Distribution of a Permian Tight Gas Field in Central Europe

5.1 Introduction

Studying the sedimentary and tectonic complexity of the Upper Permian (Rotliegend) tight gas fields in Central Europe requires integrated approaches from field-based analogue studies, laboratory analysis, seismic and well data interpretation, and structural modelling. One of the Central Europe subsurface tight gas fields that are focus of recent research is located in NW Germany, east of the Dutch Groningen gas field and is situated in the Ems Graben in ~4,200 m depth (Fig. 1.2). The tight character of this reservoir is attributed to quartz overgrowth, pressure solution and authigenic fibrous illite crystallization [1]. In general, the reservoir rocks are characterized by a heterogeneous fluvio-aeolian facies from the south-western margin of the Southern Permian Basin (SPB) that developed during the Permian Upper Rotliegend II. The prediction of sandstone reservoirs in the fluvio-aeolian deposits is challenging due to multi-phase tectonic overprinting. Further, well information is only available from present-day structural highs below the Zechstein salt, and, consequently, depositional models for the study area that cover present-day grabens are not reliable. The study area underwent a multi-tectonic overprint during the Triassic, Jurassic and Cretaceous. The original Permian structural and stratigraphic grain, including the location of Permian depocentres and associated aeolian reservoirs, was re-arranged. Most Permian fault-controlled palaeo-highs do not match the present day structural highs, while their identification is required for understanding the distribution of accommodation and its relation to sedimentary facies. Furthermore, the composition of the lowermost part of the reservoir was influenced by weathering of underlying patchy andesitic to basaltic volcanics.

The field analogue study was carried out to improve the understanding of fault control on the sedimentary facies distribution and reactive volcanic input to sedimentation in the Rotliegend tight gas field. The study took place in the northern part of the Panamint Valley (Lake Hill Basin) in Inyo County, eastern California, United States. It is located in a zone of active transtensive deformation along the

A. A. Vackiner, *Sedimentary Facies Reconstruction and Kinematic Restoration of Tight Gas Fields*, Springer Theses, DOI: 10.1007/978-3-642-36047-3_5,

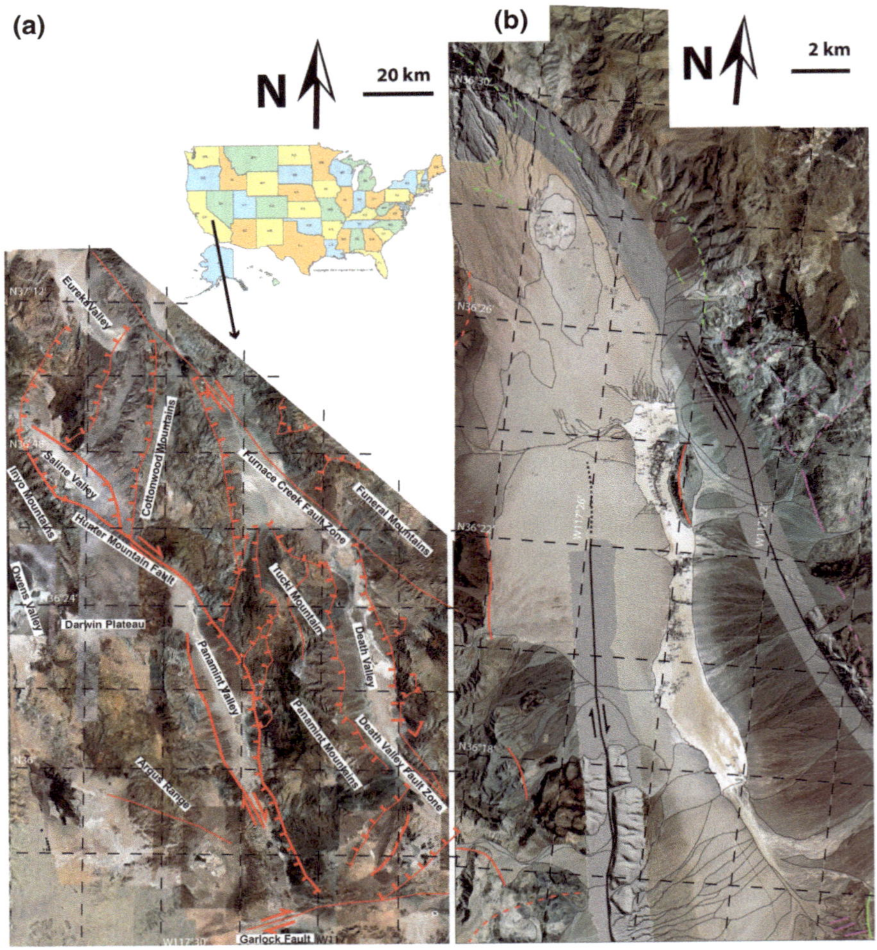

Fig. 5.1 Overview of the Panamint valley area. Satellite images copyright by google and digital globe. *Faults* and *dip* directions of faults based on field observations, LIDAR data by GeoEarthScope (*Project* GeoEarthScope Southern and Eastern California, *Target* SoCal_Panamint), and USGS digital geological maps, fault activity map of California by Jennings et al. [23] and fault map of California with location of volcanoes, thermal springs, and thermal wells by Jennings [22]. **a** Regional tectonic setting, Basin and Range province, Eastern California, Western US. Faults plotted in *red*. **b** Satellite image with LIDAR overlay and Field observation points in figure. Relative timing and description of fault activity after Jennings et al. [23] (see key to **b**). **c** Satellite image with LIDAR overlay and field observation points in figure. Facies distribution (see key to **c**). Faults plotted in *red*

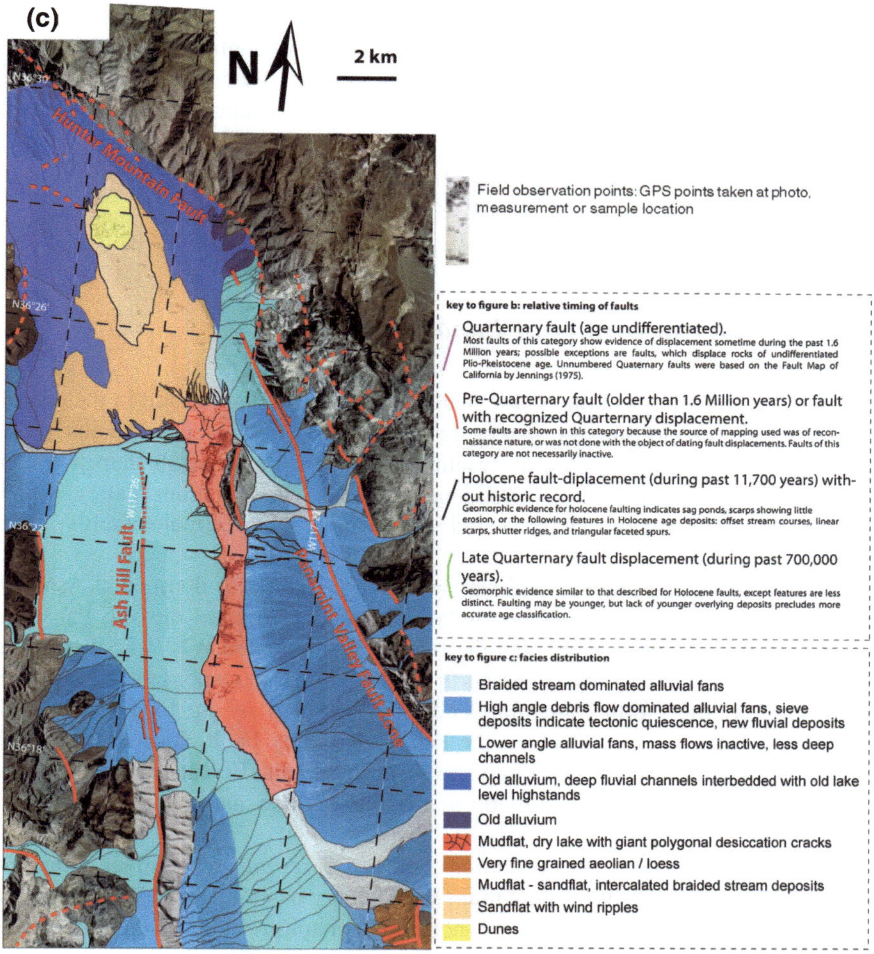

Fig. 5.1 continued

North American plate boundary [2]. The N–S-trending valley hence represents one of the active grabens that lie within the Basin and Range Province ([3]; Fig. 5.1a). The regional tectonic setting strongly influences the hydrothermal and structural characteristics of the region [4]. Large alluvial fans developed simultaneously to the recent tectonic activity and source an ephemeral shallow dry lake in the basin centre. In North Panamint Valley, an active dune field is situated on alluvial material between the dry lake and the currently active fan. This study site further provides exposed volcanics at the base of the sedimentary succession. Consequently, the general setup of tectonics, sedimentary facies and volcanic input is very similar to the subsurface study area in Germany. In addition, the field analogue provides important information about possible subsurface pathways that are open to fluid circulation. Predictive localization and control of fluid pathways is of high interest for the distribution of tight gas reservoir rock quality as enhanced fluid flow favours cementation, ultimately leading to field compartmentalization (e.g. ref. [5]).

5.2 Setting

5.2.1 Geological Setting of Panamint Valley

The Panamint Valley and the Saline Valley further north are located in the central part of the Basin and Range Province and developed as two isolated basins in one pull-apart system [6]. The general structure of the basins is a rhomb-shaped graben or half graben bordered by strike-slip faults [7, 8]. The Panamint Valley, which was the main objective of the field study, is bounded by three major fault systems that are still active (Fig. 5.1).

The N–S-trending slightly oblique right-lateral strike-slip dominated Ash Hill Fault in the W [9], the low-angle NNW-SSE-trending normal dip-slip Panamint Valley Fault Zone in the E [6], and the strike-slip dominated WNW-ESE-trending Hunter Mountain Fault Zone in the north (Fig. 5.1; [3]). In response to a long-term fault activity, the North Panamint Valley is bordered by up to 1,800 m high mountain ranges to the east along the Panamint Valley Fault Zone and by up to 800 m high hills to the west along the Ash Hill Fault. Elevations between the ephemeral dry lake, situated at ~400 m absolute altitude, and the mountain peaks reach up to 1,350 m.

The central part of the Panamint Valley represents a relatively shallow depression with an estimated maximum sedimentary basin fill of ~500 m [10]. In contrast, gravity profiles across the northern Panamint Valley acquired by the Massachusetts Institute of Technology (MIT) 1985 Field Geophysics study course and Biehler [11] indicate that the sedimentary fill does not exceed 200–300 m.

The sedimentary facies of the Panamint Valley comprises alluvial fans, a mud dominated ephemeral dry lake surface and aeolian dune, interdune, and sand flat sediments (Fig. 5.1). The ephemeral dry lake located in the centre of the valley is bounded by alluvial fans descending from the mountain ranges in the N, E and W. Towards the N, the dry lake sediments are followed by a transition of sandflat deposits and the aeolian dunes that partially cover the northern alluvial fans at ~700–830 m elevation. The present day location of the dunes is based on a northward and up-fan dune migration from a position closer to the dry lake with an average rate of 0.8 m per year [12]. Furthermore, the eastern and western alluvial fans interfinger with patchy andesitic to basaltic volcanics.

5.2.2 Subsurface Area Germany

The subsurface study area in Germany is located at the western boundary of the Ems Graben at the south-western margin of the SPB, and is characterized by a N–S-trending Zechstein salt wall situated above an asymmetric Upper Rotliegend II graben/half graben. During the Rotliegend, the SPB represented an intra-continental basin of ~1,700 km length and 300–600 km width, extending from the eastern United Kingdom to Poland and the Czech Republic ([13, 14]; Fig. 1.2). The Ems Graben in the central SPB underwent Upper Rotliegend II synsedimentary tectonics, while subsequent phases of tectonic activity, e.g. rifting in the North Sea during earliest Triassic until late Jurassic to Early Cretaceous [15] overprinted Rotliegend structural highs. The reconstructed graben structure in the study area is characterized by bounding, N–S-trending fault zones with offsets of up to 250 m in the W (Fig. 5.2, FZ-4) and up to 150 m in the E (Fig. 5.2, FZ-1). To the N, the eastern fault zone expires, and the asymmetric graben changes into a half graben (Fig. 5.2). The western fault zone bounding the graben (Fig. 5.2, FZ-4) is composed of two N–S-trending faults (FZ-4A and FZ-4B) and one orthogonal

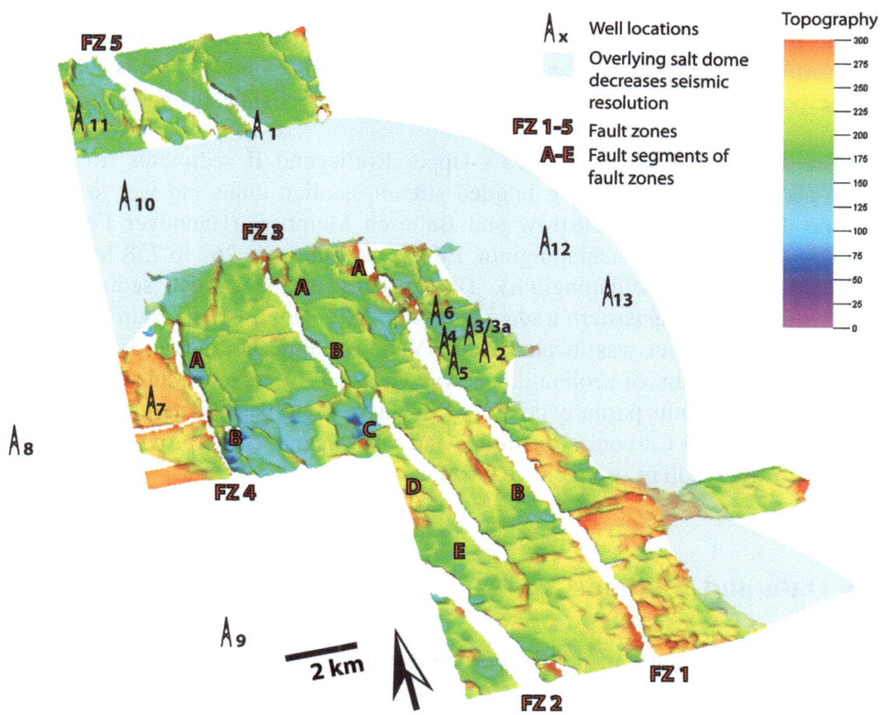

Fig. 5.2 Calculated late Upper Rotliegend II palaeo-topography derived from isopach maps with field development and exploration wells depicted

E-W fault (FZ-4C; Fig. 5.2). The palaeo-relief induced by fault activity during the deposition of the Upper Rotliegend II was estimated to a minimum of 250 m height. Coevally, a depocentre with a sediment infill of ~450 m was located on the hanging wall. The eastern zone of normal faults (Fig. 5.2, FZ-1) is composed of two major NNW-SSE-striking, westward dipping faults with cumulative vertical offsets of up to 900 m (Fig. 5.2, FZ-1A and 1B). These offsets developed in a sub-basin, which was interpreted as pull-apart basin caused by left-lateral transtension between the two major faults (Fig. 5.2, FZ-1A and 1B) and a minor eastward dipping fault (Fig. 5.2, FZ-1C). Syntectonic sedimentation in the pull-apart basin was identified by onlap geometries of Zechstein deposits in the seismic data. During the post-Rotliegend multi-phase tectonic overprint (e.g. during the Triassic), the fault zone expanded and the previously separated fault segments connected.

A third fault zone (Fig. 5.2, FZ-3) comprising a total of five major faults with different dip directions is located in the central study area. In the S, three of these normal faults dip towards the W (Fig. 5.2, FZ-3C to E), while two faults located in the N dip towards the E (Fig. 5.2, FZ-3A and 3B). The Upper Rotliegend II fault-controlled palaeo-relief with ~100–150 m height has been estimated only for the westward-dipping central fault zone. Due to the varying dip direction of faults from N to S, left-lateral strike-slip movements have been considered (Chap. 4).

The reconstructed deepest part of the graben most likely represents an area, where ephemeral dry lakes occurred in the Upper Rotliegend II. This is supported by the occurrence of a multi-order polygonal pattern visible in the seismic data [16].

Core analysis of the study area's Upper Rotliegend II sediments revealed a fluvio-aeolian origin, including braided stream, aeolian dune and wet to dry interdune deposits of the Wustrow and Bahnsen Members (Hannover Formation 260–258 Ma = early Wuchiapingium, Elbe Subgroup from 262 to 258 Ma = late Capitanium–early Wuchiapingium). The majority of the aeolian sediment was supplied by prevailing eastern trade winds [17, 18]. In contrast, the source of major fluvial sediment input was located in the Variscan hinterland towards the S [13, 19]. The preservation of aeolian dunes was governed by tectonic subsidence [20]. The sedimentary units partially cover patchy andesitic to basaltic volcanics present in core data. Local Carboniferous highs and the volcanic rocks provided essential sources for the aeolian sediment (e.g. ref. [14, 21].

5.3 Data and Methods

For a preliminary study of the large-scale tectonic setting in the Panamint Valley information from satellite images (Google Earth^{TM}; Fig. 5.1), Light Detection and Ranging (LIDAR) data provided by GeoEarthScope (Project: GeoEarthScope Southern and Eastern California, Target: So Cal_Panamint), and USGS digital geological maps ([22, 23]; Fig. 5.1) were combined. During the field study, the sedimentary facies comprising alluvial fans of different angles and braided stream systems, aeolian sediments, such as dunes, interdune and sandflat deposits, and the mudflat sediments of the ephemeral dry lake, were identified and their distribution mapped. Fault orientations were measured in the field, adapted from Jennings [22] and Jennings et al. [23] and compared to the measurements from high-resolution LIDAR data and satellite images. The composition of sediment samples from dune sands and clays of the dry lake surface, collected during the field study, were analyzed by X-ray diffractometry (XRD) and on grain compound thin-sections. The results of sediment analyses were used to determine the origin and compositional maturity of the aeolian sediments.

The data from the subsurface study area NW Germany comprises seismic interpretation of key stratigraphic horizons of the Rotliegend, faults, and the Upper Rotliegend II palaeo-topography across an area of 293 km^2 (Chap. 4). I further used well information from 14 wells including digital wire line logs from 7 wells and core data from 4 wells for the determination of sedimentary facies distribution. Dipmeter logs and one Formation Micro Imaging/Formation Micro Scanning (FMI/FMS) log were used for dip and dip directions analysis of foresets in the aeolian successions. Thin- and thick-section analysis from recovered core material on the mineral content and cement mineralogy and fluid inclusion homogenization temperatures, determined to analyze the conditions of cement genesis [24], were taken into account.

5.4 Results

The Panamint Valley was chosen as a suitable field analogue study site because of the heterogeneous fluvio-aeolian sedimentary facies distribution, the occurrence of exposed volcanics at the base of the sedimentary succession, and its synsedimentary, transtensional tectonic setting, which is of high similarity to the setting of the Upper Rotliegend II tight gas field in NW Germany. On the basis of the analogue field study in the Panamint Valley, a comparative geological model of the Upper Rotliegend II in the subsurface study area was established. In the following, the detailed sedimentary facies analysis in macro- and micro-scale and its relationship to a fault-controlled morphology is presented for both study sites.

5.4.1 Panamint Valley

Based on the field observations of the sedimentary and tectonic system, combined with available information from satellite images and geological maps, the relative timing of fault activity and the sedimentary facies distribution as well as the fault-sediment-interaction of the Panamint Valley (Fig. 5.1c) were reconstructed.

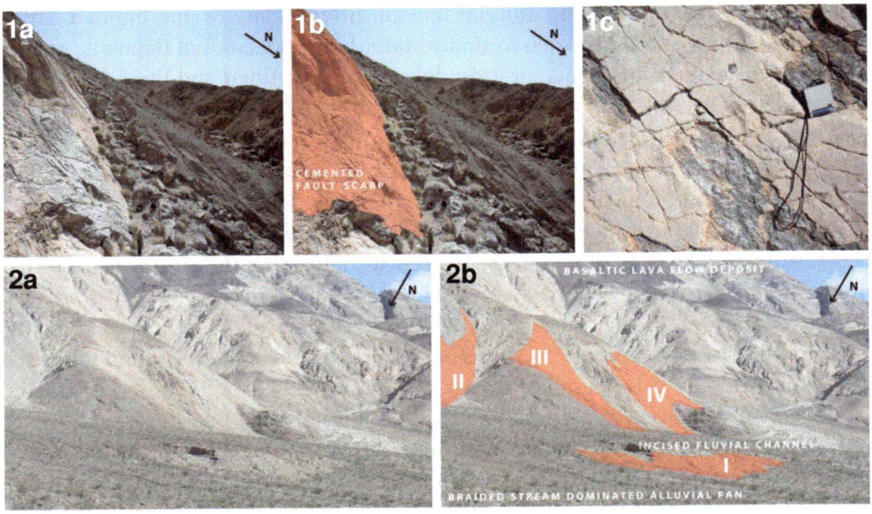

Fig. 5.3 Deformation features in the Panamint Valley. **a** Normal fault of the eastern Panamint valley fault zone, *1a* Fault scarp with cementation along offset, *1b* Fault scarp sketched in *red*, *1c* Cementation along normal fault with fault striations, cementation already grinded by fault movement. **b** View on the Panamint valley fault zone in the north-eastern part of Panamint valley, *2a* Plain Image, *2b* Image with faults sketched in *red*, *I* alluvial fan with deep incised channel offset by a normal fault, *II–IV* fault zones exhibiting fault scarps

Observed faults of the Panamint Valley are mostly intercalated and/or covered by sediments of either alluvial fans or aeolian dunes. Faults in the E further offset active channels of alluvial fans indicating recent synsedimentary fault activity (Fig. 5.3). Fault striations along cementation of visible fault scarps (Fig. 5.3) and sulphur smell suggest recent fluid circulations along open faults. Precipitations of euhedral calcite crystals of up to 0.5 cm diameter can be observed along fault scarps, but are also spread along certain stratigraphic layers, like limestone and mudstone or dolomite intervals, in the closer vicinity of faults.

To the N, E and W extensive alluvial fans descend from the footwall blocks of the outermost basin bounding faults. The alluvial fans cover the major faults, e.g. the Ash Hill Fault in the W and the Panamint Valley Fault Zone in the E, which, in turn, interact with the active channel network of the fan deposits (Fig. 5.3). In the central part of the basin the alluvial fans partly cover and drain the dry lake surface. The lack of desert varnish implies recent sediment transport along all alluvial fans.

Two types of alluvial fan systems that can be divided by differing slope angles in the north-eastern and the north-western basin can be differentiated. Braided stream dominated alluvial fans are mainly located in the NW of the Panamint Valley e.g. along the Northern Ash Hill Fault Zone, and show slopes of $\sim 4°$. They are characterized by deep and wide incised channels. The braided streams show coarse grained gravel bars without internal geometries and fine to medium grained

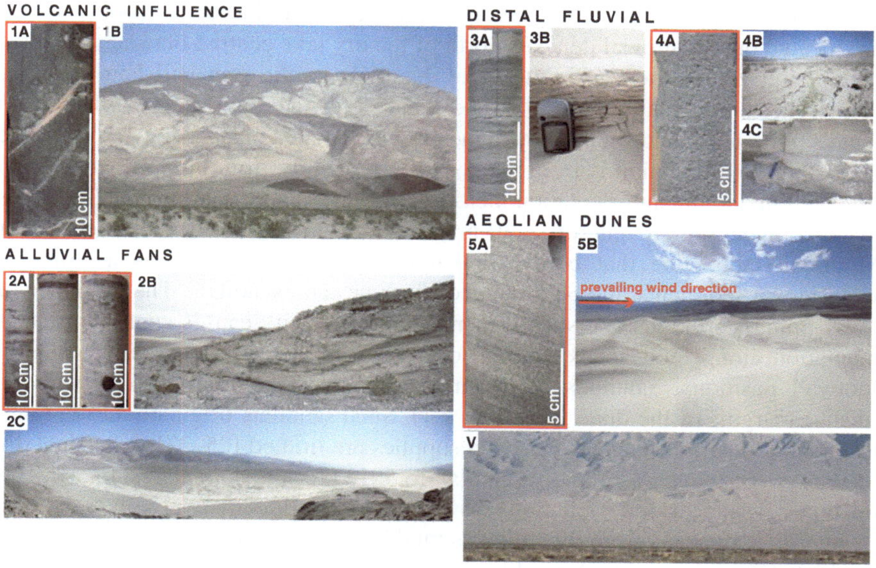

Fig. 5.4 Comparison of core material from the tight gas field in NW germany (indicated by *Arabic numbers*) to field observations from the Panamint valley (indicated by *Latin numbers*) *1* Basaltic to andesitic lava flow *I* Basaltic to andesitic lava flow in Panamint valley, present on the footwall and on the hangingwall (intercalated into alluvial fan) of the shallow detachment Panamint valley fault zone. *2* Braided stream alluvial fans with gravel bar and cross-stratified channel deposits. The core depicted in the middle might already represent first dune successions. *II* View on main fanhead channel in the background, passing into a distributive network of channels and distributive network of sand-filled shallow channels, sourced from alluvial fans and finally passing into the dry lake on the right. *3* Wet sandflat deposit with intercalated clay. *III* Wet sandflat deposit as interdune environment. *4* fluvial sand-filled channel with clay rip-up clasts. *IV* Sand-filled channel. *5* Barchanoid dune deposits with changing dipping angles between the cross-stratifications. *V* Barchanoid to aklé dune deposits and Panamint Dunes being located on incised channels of older alluvium

intra-channel deposits with ε-cross-stratification. In the north-eastern Panamint Valley, debris flow dominated alluvial fans with slope angles of up to 10° (Fig. 5.4) characterize the hanging wall sedimentology of the Panamint Valley Fault Zone. At the intersection point of the fan trench with the general alluvial fan surface, the confined flow along fan-head channels changes into a distributive flow along a network of sand-filled shallow channels (Fig. 5.4). Comparing the Panamint Valley Fault Zone in the E and the Ash Hill Fault in the W of the Panamint Valley shows that (1) the occurrence of braided stream dominated alluvial fans is associated with shallower topography and higher erosion rates of Tertiary to Quaternary volcanics and Palaeozoic shales and carbonates, and (2) the occurrence of debris-flow-dominated alluvial fans is associated with higher relief and less erosion of Palaeozoic carbonates, Tertiary granitoids and Tertiary to Quaternary volcanics (Fig. 5.1).

The Panamint Dunes are prevailing barchanoid and aklé dunes (Fig. 5.4) and are located on old alluvial deposits north of the dry lake centred between the three major fault zones (the Panamint Valley Fault Zone, the Ash Hill Fault and the Hunter Mountain Fault; Fig. 5.1). The dune field, including dry sandflats with wind ripples north and south of the dunes, covers an area of $\sim 2.1 \times 4.5$ km^2. The dominant wind direction is from the S and infrequently from the NE, which is in line with the dominant orientation measured for dune and ripple ridges. The dunes reach a maximum height of ~ 30 m. The fault-induced relief of the Hunter Mountain fault zone that borders Panamint Valley to the N acts as trap for wind-blown sediments that build up aeolian dunes and sandflats. The existence of overlapping step faults and synsedimentary fault activity causes continuous sedimentation along fault zones leading to easier up-wind access of dune sands to footwall positions. Sandflats occur on the up-wind (proximal) and down-wind (distal) margins of the dune field (Fig. 5.1). Such sandflats originate during long-lasting strong winds and/or when sand supplies are limited [25–27]. A gradual dip increase from 0 to 5° for the dry sandflat, to 5–15° for the dune bases, to >15° for the dunes was measured. Deflation lags accompany or directly overly the sandflats.

Interdune deposits were identified as muddy to fine grained sandy damp to wet sandflats (Figs. 5.4 and 5.5), mudflats, lake margins (Figs. 5.4 and 5.5) and ponds (Figs. 5.4 and 5.5). Dune marginal deposits (Figs. 5.4 and 5.5) are composed of fine grained sand with interfingering clay. In many cases a progradation of the lee side of aeolian dunes into the ponds or lakes was observed. Ripple lamination along marginal lake or pond deposits is common.

The damp and wet sandflat deposits (Figs. 5.4 and 5.5) were identified as siltstones or very fine to fine grained poorly sorted sandstones. I assume that they

Fig. 5.5 Close lateral changes in aeolian sediment *1–4* core material of the subsurface area in NW Germany, *1* damp sandflat deposit, *2* aeolian dune base deposit, *3* pond deposit, *4* lake or pond margin deposit. *I–IV* closely spaced lateral changes with dune successions (dry sandflat—aeolian dune), pond and damp-wet sandflat. Arabic numbers of the sedimentary facies in the cores correlate with roman numbers of the sedimentary facies in the modern analogue

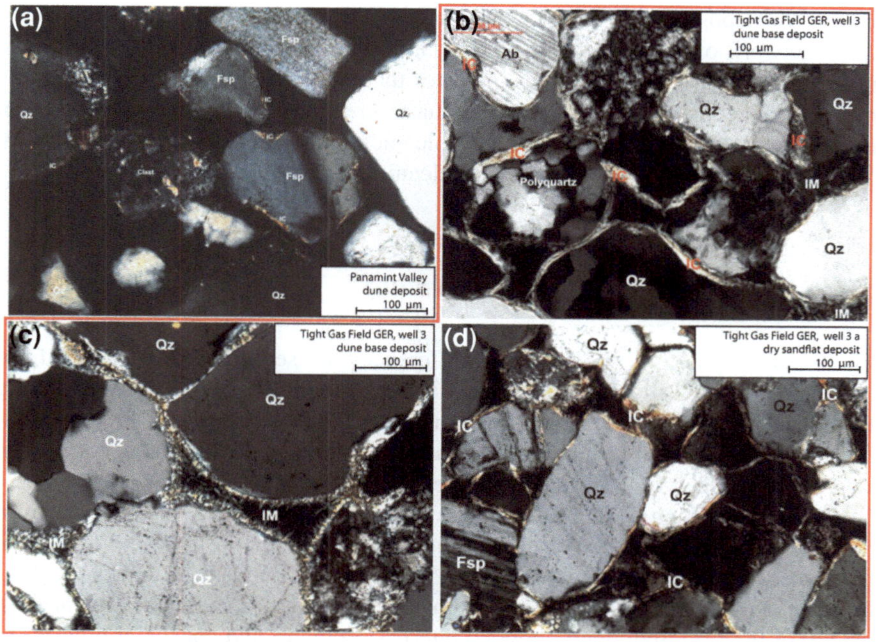

Fig. 5.6 **a** Thin-section picture of recent dune sample of the Panamint valley dunes. **b–d** Thin-section pictures of samples from the Wells 3 and 3a of the NW German tight gas field

deposited under influence of the shallow groundwater table at ~3 m depth and combined with ephemeral flooding [28, 29]. The aeolian sedimentation is further influenced by saline groundwater and by adhesion of wind-transported aeolian grains to a damp sediment surface. Small current ripples or horizontal laminations are common. From the XRD measurements of the collected dune sand samples quartz, feldspar and calcite as main components were identified. Accessories are illite/muscovite, chlorite, kaolinite, and amphibole (Fig. 5.6). Thin section analysis further reveals high percentages of lithoclasts. Based on the classification by Pettijohn [30] the sand composition is litharenitic. This composition indicates that primary local sediment sources of the sand dunes are the Quaternary alluvial fans, the Tertiary volcanics, the Mesozoic to Tertiary granitoids, and the Palaeozoic dolomites to limestones. XRD measurements [16] of samples from the lake surface revealed that the main components are: dolomite, quartz, illite/muscovite, calcite and feldspar. Accessory minerals are chlorite, kaolinite, hematite, thernadite and amphiboles. Swelling clays could be identified by glycol dehydration of the clay fraction.

Patchy basaltic to andesitic volcanics cover large parts of the mountain ranges surrounding the Panamint Valley (Figs. 5.1 and 5.4). These relatively thin layers were formed by lava flows [31] that followed the ancient topography, and therefore are found also along the major channel systems on the hanging walls of faults.

Lava flow deposits were also observed in the alluvial fans, rinsed by the distributive network of braided channels (Fig. 5.4). The high amount of volcaniclastic material found in the alluvial fans indicates that the volcanics served as one of the major local sediment sources. Further, quartz grains from the dune succession on top of the alluvial fans and in front the volcanics show chlorite coatings, which are assumed to originate from weathered volcanic material.

5.4.2 Subsurface Study Area Germany

The seismic analysis showed that differences in sediment thicknesses from hanging- to footwalls revealed synsedimentary activity of the graben during Rotliegend II sediment accumulation (Chap. 4). The complexity of the fault-sediment-interaction further increased towards the development of fault zones, e.g. step faults induced by transtensional stresses.

The interpretation of the sediment content in the lower parts of core data from Wells 2, 3 and 3a showed coarse grained gravel bar deposits without internal geometries and fine to medium grained intra-channel deposits with ε-cross-stratification (Fig. 5.4.2a) that originate from braided stream dominated alluvial fans. On top of the fans, sediments show no internal geometry and consist of breccias to conglomerates with angular rounded grains up to 5 cm diameter. This facies is interpreted as flashflood deposits originating from hyper-concentrated gravitational mass flows (Fig. 5.4.2a, right core photo). Frequently interbedded sandstones with bimodal fine and medium grain sizes and cross-stratification are either remains of aeolian dune deposits or fluvial reworked aeolian depositional systems (Fig. 5.4.2a).

On top of the alluvial intervals, the sedimentary facies changes towards an aeolian succession with dry (dry sandflat, aeolian dune base, aeolian dune) to wet (interdune mudflat, interdune pond or lake, fluvial and pond or lake margin) deposits (Fig. 5.5). The 'shale line' in Gamma Ray (GR) logs implies that the (fluvio-) aeolian-dominated intervals reach thicknesses of up to 150 m in wells at the footwall of FZ-1, and only up to 50 m at the footwall of FZ-4 (Fig. 5.7). The evaluation of dipmeter logs showed that the footwall of FZ-4 is mainly composed of low-dipping sandflat deposits and wet deposits producing an irregular log signature. FMI/FMS and dipmeter log analysis in sandstone dominated well succession of the FZ-1 footwall indicate higher dips in dune dominated areas with main dip direction to the W, partly superimposed by an irregular pattern of damp sandflat deposits (Fig. 5.7). Measured dip direction to NE to NNE are in line with the prevailing wind direction [17, 18].

The interpretation of core material reveals that individual dune sets (dune base + dune) show maximum thicknesses of 3 m, and are of barchanoid or small transverse type or barchanoid with amalgamated dune ridges (aklé dunes; Fig. 5.4.v). As a result of syn- and post-sedimentary erosion of aeolian accumulation, the initial dune set thicknesses are difficult to determine. However, the

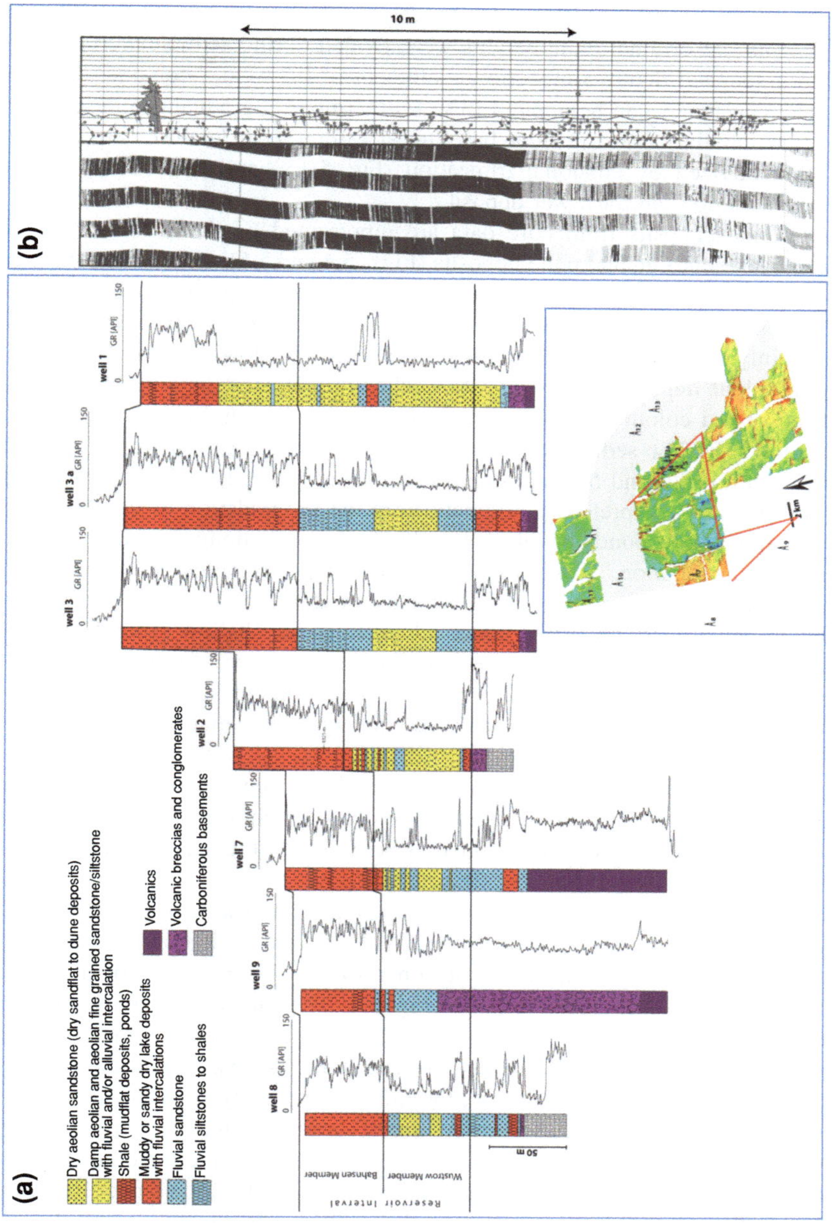

Fig. 5.7 **a** GR logs and simplified lithology and facies interpretation of wells of the subsurface study area in East Frisia, NW Germany. Plot flattened on Base Reservoir (Base Wustrow Member). Note the thickness variations over relatively closely spaced lateral areas. **b** Example of a log interval from the combined FMI/FMS and dipmeter log of Well 3a

maximum initial dune height was estimated to ±20 m, assuming that (1) about 60 % from the original dune heights were eroded during sedimentation [32] and (2) an associated depth compaction coefficient of 0.27 km^{-1} for fine to medium grained sandstones can be applied [33]. Furthermore, a gradual dip increase from <5° (dry sandflat) to 5–15° (dune base) to >15° (dune) can be observed. Dune tops are characterized by erosional truncation (Fig. 5.4.v) and overlain by the next dune, a deflation lag, or a wet deposit.

Interdune deposits from core data are represented by damp to wet sandflat, mudflat, lake margin, and pond deposits (Figs. 5.4 and 5.5). Aeolian mudflats with post-sedimentary structures, such as convolute bedding and ball-and-pillow-structures, are the most significant interdune deposits. Their deposition took place in a mainly subaquateous milieu with adhesive sands blown in [34]. The pond or lake interdune deposits consist of 100 % clay, do not show any internal structure, and are of red colour due to the absence of the reducing effect of hydrocarbon migration [35]. The sediments in the transition from dune to interdune pond/lake deposits (Figs. 5.4 and 5.5) are composed of fine grained sand with intercalated clays. They are interpreted to originate from the progradation of the lee sides of aeolian dunes into ponds or lakes. Ripple laminations along marginal pond/lake deposits are common. Damp to wet sandflat deposits (Figs. 5.4. and 5.5) are siltstones or very fine to fine grained, poorly sorted sandstones with irregular wavy bedding, which were deposited under influence of the shallow groundwater table and ephemeral flooding [34]. The damp sediments were further reworked by precipitation and dissolution of salt below the sediment surface, as well as by adhesion of wind transported grains, as indicated by small current ripples or horizontal laminations.

Thin-section analysis of the aeolian dune and interdune deposits proved quartz, feldspar, lithoclasts and clay minerals comprising chlorite and illite as main components. Sandstone classification after Pettijohn [30] reveals that the sandstone is a litharenite.

At the base of the sedimentary succession, patchy andesitic to basaltic volcanic lava flows of up to 120 m thickness (e.g. Well 7) were drilled on footwalls of the western (FZ-4) and the eastern fault zone (FZ-1). An example for a vesicular textured lava flow deposit is shown in Fig. 5.4. Their tops are heavily brecciated due to reworking by e.g. gravitational mass transport of sediments on top. Therefore, a high amount of volcaniclastic material is observed in the sediments overlying the volcanics. In addition, the thin-section analysis of the aeolian dune deposits above the volcanics showed quartz grains coated by early-diagenetic chlorite that originates from chemical weathering of volcanic material (Fig. 5.6).

Due to the fact that the seismic resolution at the depth of the Upper Rotliegend II target interval in NW Germany only allows basic interpretations and the wells provide only punctual information, conclusions about lateral continuity or distribution of sedimentary facies or volcanics can only be roughly inter- and extrapolated. During the core interpretation and the work on isopach maps, uncertainties of the relative positions of sediment bodies in relation to palaeo-topography emerged. I combined the observations from core material and log interpretation

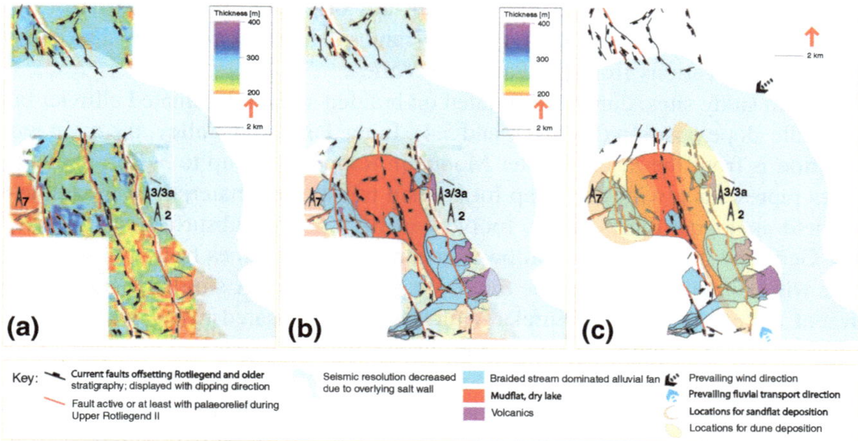

Fig. 5.8 **a** Upper Rotliegend II thickness map of the subsurface tight gas study area in NW Germany. **b** Facies distribution and information of the Panamint Valley applied to the German study area with respect to well log and core interpretations and potential facies distribution on seismic and isopach maps plotted onto Upper Rotliegend II isopach map. **c** Wind and fluvial transport directions indicate potential accommodation for aeolian dunes and sandflats

with results of seismic interpretation and 3-D surface tracking and summarized them to a model of relative timing of fault activity and sedimentary facies distribution (Fig. 5.8).

5.5 Discussion

From the observations in the field and the subsurface study, I propose that the Panamint Valley with its complex synsedimentary, transtensional fault zone activity and related sediment facies distribution represents a field analogue for the subsurface study site in Germany. In the following, the model of the Panamint Valley's fault controlled topography as a controlling factor of the sedimentary facies distribution, also considering wind direction and sediment source rocks, was used and applied to the subsurface study site (Fig. 5.8). Because the elevation between the ephemeral dry lake and the mountain peaks in Panamint Valley is 350 m in the W and 1,350 m in the E, I limit the application of the analogue study and focus on the western flank's topography for the comparison with the subsurface study site and its reconstructed palaeo-topography.

In the Panamint Valley, alluvial fans sourced from the north-western range are braided stream dominated alluvial fans (Fig. 5.1). Because of the similarity between the topographic gradient of the modern analogue and the reconstructed palaeo-topography of the subsurface study site, I assume that mainly braided stream dominated alluvial fans developed in the German study site as well. The palaeo-elevations, from which the alluvial fans originate, mainly consist of Carboniferous carbonates and Rotliegend volcanics. This observation is also in-line with the observations from the Panamint Valley.

In both study sites, dunes are situated on braided stream dominated alluvial fans in middle slope position (Figs. 5.1 and 5.4). In the Panamint Valley, the main wind direction is from the S; the Hunter Mountain Range rising up to 370 m above the dunes represents a windward trap for aeolian transported material. Comparable to the field analogue, dunes on the footwall of FZ-1 in the subsurface study area in NW Germany developed as shallow barchanoid to aklé dunes from a unimodal E/ENE wind direction in a lee side trap (Figs. 5.4 and 5.5). A similar lee side trap is present for example in the Eureka Valley, which is located to the N of the Panamint Valley. The Eureka Valley dunes have irregular ridges and are up to 208 m high [36].

Well logs from the German study area (Fig. 5.7) from the footwall of FZ-4 reveal the occurrence of sandflats deposited on top of fluvial sandstones and conglomerates with underlying monomict volcanic breccias to conglomerates and volcanics. Based on the facies distribution map of the Panamint Valley and the E/ENE wind direction, I expect that the fault zone represented a windward trap, with a higher amount of conglomerates and sand accumulation present in hangingwall position. However, the dunes of the Panamint Valley and the subsurface site show a high similarity of dune heights (i.e., 20–30 m), the amount of interdune deposits, and the dune sediment composition analyzed from thin-sections. Furthermore, dune bases at both study sites migrate into interdune deposits causing post-sedimentary deformation structures in alternating sand, silt and clay deposits (Figs. 5.4 and 5.5).

The ephemeral dry lake in the Panamint Valley (Fig. 5.1), which is characterized by the occurrence of huge desiccation polygons, is located in the deepest basin area. The deepest Rotliegend basin area of the asymmetric graben to half-graben is located on the hangingwall of FZ-4 of the German subsurface study site. I interpret that it represents a trap for dry to ephemeral lakes, or was at least exposed to the influence of the groundwater table. This is supported by the multi-order polygonal pattern in this part of the basin that is very similar in shape and size to the polygonal pattern of huge desiccation cracks on the dry lake surface in Panamint Valley (Antrett et al., in press).

In both study sites patchy andesitic to basaltic volcanic lava flow deposits occur in footwall position (Fig. 5.4). In the Panamint Valley, these volcanics are also intercalated with the alluvial fans in hangingwall position. Consequently, the hangingwall volcanics are mechanically weathered by the alluvial fans' braided channel system and serve as active sediment source (Fig. 5.4). Due to the similar vesicular, lava-flow-typical texture, I propose the occurrence of volcanics in undrilled hangingwall positions also for the subsurface study area. The input of altered volcanic material from the footwall into the deeper part of the basin provides a source of reactive Al^{3+} and Si^{4+} and supports the early and abundant diagenetic development of alumosilicate minerals [37]. Swelling clays (e.g. Smectite) therefore most likely originate from the weathering of the volcanics

[38]. Thin-section analysis of grain samples of both the Panamint Dunes and the Upper Rotliegend II dune deposits reveal discontinuous coating of quartz and feldspar grains by detrital or synsedimentary early diagenetic chlorite and illite (Fig. 5.6), which is also provided by the weathering of volcanic material and transported via the alluvial fan channel system. Discontinuous, mechanical abrasion of clay coatings around quartz grains are observed for both study sites. As this is associated with active dune migration in the Panamint Valley, I anticipate active dune sediment transport being responsible for coating abrasion in the German subsurface study site, too (Fig. 5.6).

5.6 Conclusions

1. The Panamint Valley's heterogeneous sedimentary facies distribution comprises two different types of alluvial fans, sand dunes and sandflats. The distribution of these facies is controlled by topography and synsedimentary faults, the local sediment sources, and the prevailing wind direction. The abrasion of quartz grain coatings, dune types and sizes, the presence or absence of desert varnish, and the incision depth of alluvial fan channels can be used as proxies for estimating the sediment dynamics.
2. Core data analysis (Sect. 3.2) of the sedimentary facies from the subsurface tight gas reservoir in Germany reveals braided stream dominated alluvial fan deposits, stacked sand dune, sandflat and interdune deposits. Sediment thicknesses and facies significantly vary between cores across major fault zones.
3. The presence of patchy basaltic to andesitic volcanic lava flow deposits in both study sites strongly influenced the sediment composition and supplied clays into the sedimentary systems.
4. A model of topography, synsedimentary faults, sediment sources, and wind direction as the key controlling factors of the sediment facies distribution in the analogue study site was developed and compared to the subsurface site reconstructed to the Upper Rotliegend II setting prior to multi-phase tectonic overprinting. The comparison shows that the Panamint Valley represents a well-suited modern analogue for the German subsurface tight gas reservoir. Especially the localization of the aeolian sandstones, comprising dune and sandflat deposition controlled by fault-induced topography that acts as windward and lee side trap for aeolian sand, is of high similarity (Fig. 5.9). Furthermore, the field analogue observations concerning abrasion of quartz grain coatings, dune types and sizes, the presence or absence of desert varnish, and the incision depth of alluvial fan channels was transposed to reconstruct the sediment dynamics of the tight gas reservoir during the Upper Rotliegend II.
5. In a general context, the study shows that a well-suited field analogue study enables (1) a detailed interpretation, interpolation, extrapolation, and prediction of a sedimentary facies distribution for only spatially limited core data across

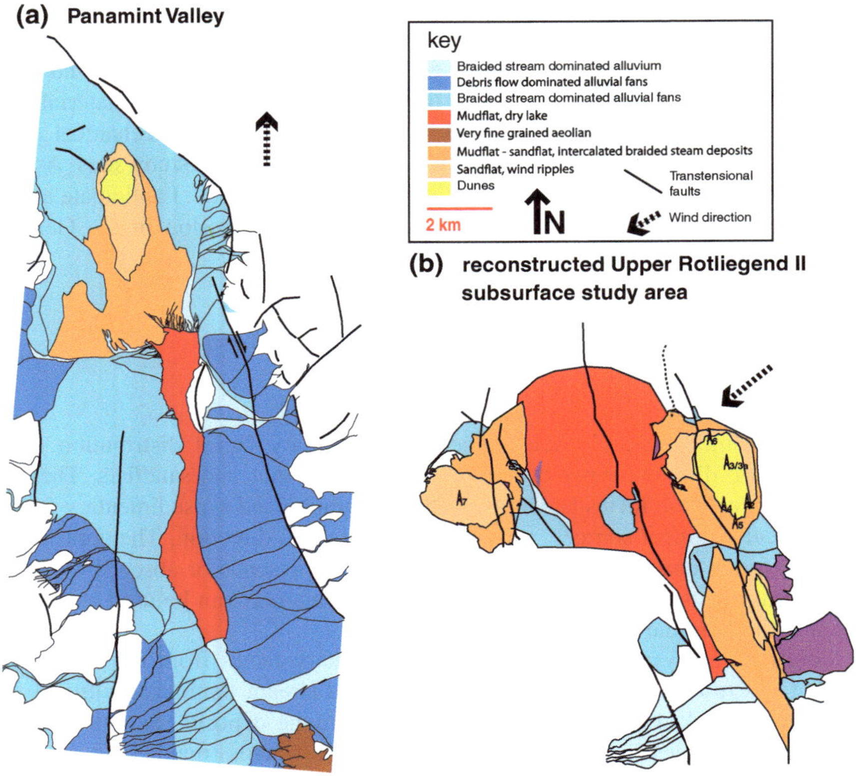

Fig. 5.9 Comparison between the simplified sedimentary facies distribution of Panamint Valley on the *left* (**a**), of the German subsurface study area on the *right* (**b**)

larger areas, (2) a transfer of key mechanisms of sedimentary facies distribution to subsurface data using fault interpretation, retro-deformation and palaeo-relief reconstruction, and (3) the reconstruction of sediment dynamics in tectonically overprinted subsurface areas by detailed observations of modern field analogues.

References

1. Gaupp R, Solms M (2005) Palaeo oil- and gasfields in the Rotliegend of the North German basin: effects upon hydrocarbon reservoir quality (Paläo-Öl- und Gasfelder im Rotliegenden des Norddeutschen Beckens: Wirkungen der KW-Migration auf die Speicherqualitäts-Entwicklung.). In: Gaupp R (ed) DGMK research Report 593: Tight Gas Reservoirs—Natural Gas for the Future: DGMK Celle, p 242

2. Reheis MC, Sawyer TL (1997) Late Cenozoic history and slip rates of the fish lake valley, emigrant peak, and deep springs fault zones, Nevada and California. Geol Soc Am Bull 109:280–299
3. Smith SU (1976) Late-quarternary pluvial and tectonic history of panamint valley, Inyo and San Bernardino Countries. Ph.D. thesis, California Institute of Technology, p 295
4. Jayko AS, Forester RM, Kaufman DS, Phillips FM, Yount JC, McGeehin J, Mahan SA (2008) Late Pleistocene lakes and wetlands, Panamint Valley, Inyo County, California. Geol Soc Am Spec Pap 439:151–184
5. de Medeiros WE, do Nascimento AD, Antunes AF, de Sá EFJ, Neto FFL (2007) Spatial pressure compartmentalization in faulted reservoirs as a consequence of fault connectivity: a fluid flow modelling perspective, Xaréu oil field, NE Brazil. Petrol Geosci 13:341–352
6. Burchfield BC, Hodges KV, Royden LH (1987) Geology of Panamint Valley—Saline Valley pull-apart system, California: Palinspastic evidence for low-angle geometry of a Neogene range bounding fault. J Geophys Res 92(B10):10422–10426
7. Aydin A, Nur A (1985) The types and role of stepovers in strike-slip tectonics (in Strike-slip deformation, basin formation, and sedimentation): Special publication—society of economic palaeontologists and mineralogists, vol 37. pp 35–44
8. Price N, Cosgrove J (1990) Analysis of geological structures. Cambridge University Press, Cambridge, p 520
9. Densmore AL, Anderson RS (1997) Tectonic geomorphology of the Ash Hill fault, Panamint valley, California. Basin Res 9:53–63
10. Blakely RJ, Jachens RC, Calzia JP, Langenheim VE (1999) Cenozoic basins of the Death Valley extended terrane as reflected in regional-scale gravity anomalies. In: Wright LA, Troxel BW (eds) Cenozoic basins of the Death Valley region: Special Paper—Geological Society of America, vol 333. pp 1–16
11. Biehler S, MIT Geophysics Field Course (1987) A geophysical investigation of the Northern Panamint valley, Inyo County, California: evidence of possible low-angle faulting at shallow depth in the crust. J Geophys Res 92(B10):10427–10441
12. Prestud Anderson S, Anderson RS (1990) Debris-flow benches: dune-contact deposits record palaeo-sand dune positions in north Panamint valley, Inyo County, California. Geology 18:524–527
13. Plein E (1993) Bemerkungen zum Ablauf der palaeogeographischen Entwicklung im Stefan und Rotliegend des Norddeutschen Beckens. Observations on Stephanian and Rotliegendes palaeogeography in the North German Basin. In: Zur Geologie und Kohlenwasserstoff-Fuehrung des Perm im Ostteil der Norddeutschen Senke. Geology and hydrocarbon potential of the Permian rocks of the eastern North German Basin: Geologisches Jahrbuch. Reihe A: Allgemeine und Regionale Geologie BR Deutschland und Nachbargebiete, Tektonik, Stratigraphie, Palaeontologie, vol 131. pp 99–116
14. McCann T (1998) The rotliegend of the NE German basin: background and prospectivity. Petrol Geosci 4:17–27
15. Ziegler PA (1990) Geological atlas of Western and Central Europe, 2nd edn. Shell, The Hague, p 239
16. Antrett P, Vackiner AA, Kukla P, Klitzsch N, Stollhofen H (2012) Impact of arid surface megacracks on hydrocarbon reservoir properties. AAPG Bulletin 96(7):1279–1299
17. Gast RE (1988) Rifting im Rotliegenden Niedersachsens, Rifting in the Rotliegendes of Lower Saxony: Die Geowissenschaften Weinheim 6(4):115–122
18. Rieke H, Kossow D, McCann T, Krawczyk C (2001) Tectono-sedimentary evolution of the northernmost margin of the NE German basin between uppermost Carboniferous and Late Permian (Rotliegend). Geol J 36(1):19–38
19. Glennie KW (1990) Introduction to the petroleum geology of the North Sea, vol 3. Wiley, New York, p 416
20. Kocurek G (2003) Limits on extreme eolian systems; Sahara of Mauritania and jurassic Navajo Sandstone examples. In: Chan MA, Archer AW (eds) Extreme depositional

environments; mega end members in geologic time: Geological Society of America special paper, vol 370. pp 43–52
21. Glennie KW (1990a) Rotliegend sediment distribution; a result of late Carboniferous movements. In: Hardman RFP, Brooks J (eds) Proceedings of tectonic events responsible for Britain's oil and gas reserves, vol 55. Geological Society of London, Special Publications, pp 127–138
22. Jennings CW (1975) Fault map of California with location of volcanoes, thermal springs, and thermal wells: California division of mines and geology geologic data map No. 1, scale 1:750,000, 1 sheet
23. Jennings CW, Bryant WA, Saucedo G (2010) Fault activity map of California. California geological survey 150th anniversary: California geologic data Map series map No 6, scale 1:750,000, 1 sheet
24. Havenith VMJ, Meyer FM, Sindern S (2010) Diagenetic evolution of a tight gas field in NW Germany. DGMK/ÖGEW-Frühjahrstagung 2010, Fachbereich Aufsuchung und Gewinnung, Celle
25. Fryberger SG, Ahlbrand TS, Andrews S (1979) Origin, sedimentary features and significance of low-angle eolian 'sand sheet' deposits. Great sand dunes national monument and vicinity, Colorado. J Sediment Petrol 49:733–746
26. Fryberger SG, Al-Sari AM, Clisham TJ (1983) Eolian dune, interdune, sand sheet, and siliciclastic sabkha sediments of an offshore prograding sand sea, Dhahran area, Saudi Arabia. AAPG Bulletin 67:280–312
27. Kocurek G (1988) First-order and super bounding surfaces in eolian sequences—bounding surfaces revisited. Sed Geol 56(1–4):193–206
28. Fryberger SG, Schenk CJ, Krystinik LF (1988) Stokes surfaces and the effects of near-surface groundwater-table on aeolian deposition. Sedimentology 35(1):21–41
29. Meadows NS, Beach A (1993) Structural and climatic controls on facies distribution in a mixed fluvial and aeolian reservoir; the Triassic Sherwood Sandstone in the Irish Sea. In: North CP, Prosser DJ (eds) Characterization of fluvial and aeolian reservoirs. Geological Society of London, Special Publication, vol 73. pp 247–264
30. Pettijohn FJ (1963) Chemical composition of sandstones; excluding carbonate and volcanic sands, Chapter S. In: Data of geochemistry, vol 6. United States Geological Survey Professional Paper, pp 1–21
31. Andrew JE, Walker JD (2009) Reconstructing late cenozoic deformation in central Panamint Valley, California: evolution of slip partitioning in the walker lane. Geosphere 5(3):172–198
32. Allen PA, Allen JR (1990) Basin analysis: principles and applications. Blackwell Sciences, USA, p 451
33. Sclater JG Christie PAF (1980) Continental stretching: an explanation of the post-mid cretaceous subsidence of the central North Sea basin: J Geophy Res 85:3711–3739
34. George GT, Berry JK (1993) A new palaeogeographic and depositional model for the Upper Rotliegend of the UK sector of the Southern North sea. In: North CP, Prosser DJ (eds) Characterization of fluvial and aeolian reservoirs, vol 73. Geological Society of London, Special Publication, pp 291–319
35. Chan MA, Parry WT, Bowman JR (2000) Diagenetic Hematite and Manganese Oxides and fault-related fluid flow in jurassic sandstones, Southeastern Utah. AAPG Bull 84:1281–1310
36. Norris RM (1987) Eureka Valley sand dunes. In: Hall CA, Doyle-Jones V (eds) Plant biology of Eastern California. Natural history of the white-Inyo range: Symposium, Vol 2. p 207–211
37. Jeans CV, Wray DS, Merriman RJ, Fisher MJ (2000) Volcanogenic clays in jurassic and cretaceous strata of England and the North sea basin. Clay Miner 35:22–55
38. Roen JB, Hosterman JW (1982) Misuse of the term 'bentonite' for ash beds of Devonian age in the Appalachian basin. Geol Soc Am Bull 93:921–925

Chapter 6
Integrating Salt Kinematics and Diagenesis in a Tight Gas Field: A Case Study from the Upper Rotliegend in East Frisia

6.1 Introduction

In this chapter, a sequential retro-deformation of the tectonic history of an Upper Rotliegend II tight gas field in NW Germany (Fig. 6.1), and its linkage to sequential phases of diagenetic processes is presented. During the deposition of the Upper Rotliegend II reservoir rocks, the study area was located at the southern margin of the Southern Permian Basin (SPB), a sub-basin of the Central European Basin system, experiencing a complex burial history (e.g. [1, 2]). Table 6.1 provides a summary of lithologies and related depositional environments, as well as contemporaneous tectonism and climatic variations that affected the study area through time.

The applied stepwise retro-deformation aims to provide discrimination between salt rise mechanisms, the timing of salt deformation, and their relation to regional tectonic events. The retro-deformation technique thereby considers sedimentation, decompaction, fault related deformation, salt movement, thermal subsidence, and isostasy [3]. The reconstructed timing of tectonics and salt kinematics is then compared to key diagenetic processes detected from petrographic inspection of core plug samples, integrated with microthermometry analyses of fluid inclusions in cement-building mineral phases.

Retro-deformation of extensional basins involves the progressive removal of sediment loads considering isostatic effects and sediment decompaction [3]. Such section restoration techniques were initially developed to restore effects of compressional tectonics [4], but were later also applied to areas characterized by extensional tectonics [5]. Hossack and McGuinnes [6], Rowan [7], Bishop et al. [8], and Buchanan et al. [9] successfully applied this method to regions, which have undergone salt tectonics. In the NW German study area, local tectonics and synsedimentary structures are strongly influenced by the rise and fall of Late Permian Zechstein salt diapirs and pillows. Due to the behaviour of salt as viscous fluid throughout geological time spans (e.g. from studies on the complexity of salt kinematics: [10, 11]; from analogue modelling studies: [12–14]; and from numerical modelling studies: [15–17]), a retro-deformation in areas influenced by

A. A. Vackiner, *Sedimentary Facies Reconstruction and Kinematic Restoration of Tight Gas Fields*, Springer Theses, DOI: 10.1007/978-3-642-36047-3_6,

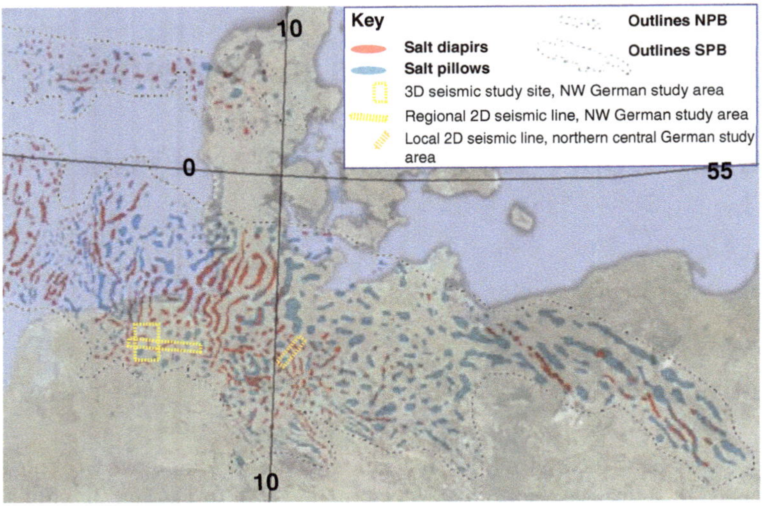

Fig. 6.1 Map of the study area in context with locations of Zechstein salt diapirs and pillows in the Northern and Southern Permian Basin (after Lokhorst [62]). *NPB* = Northern Permian Basin, *SPB* = Southern Permian Basin. Basin outlines modified from Ziegler [61] and Legler [46]

salt tectonics is challenging and has several limitations. However, in this study I present the retro-deformation of an area, in which multi-disciplinary integrated research has been undertaken, including fluid inclusion measurements [18, 19] and isopach map analysis, which provide valuable information about sediment loads and burial history through time. The uncertainties of the retro-deformation are thus extenuated.

6.2 Data and Methods

The study area is located in East Frisia, NW Germany, covered by a 293 km^2 3D seismic volume in pre-stack depth migration (PSDM) and post-stack time migration (PSTM; Figs. 6.2, 6.3). The seismic data set contains a regional, 100 km long, approximately W-E-oriented, PSTM 2D seismic line that crosses the southern part of the 3D seismic survey (Fig. 6.3). The regional 2D seismic line was depth converted using a sonic log calibrated velocity model in order to connect it to the PSDM data and wells. Seismic multi-attribute (variance, chaos) analysis provided information about the palaeo-sedimentary and palaeo-geomorphologic settings. Furthermore, log data of 14 wells (FMI/FMS logs are only available for Well 3a) located on structural highs below the Zechstein salt were integrated. Core material of three wells (2, 3, 3a) was analyzed with a focus on the Upper Rotliegend II. Macroscopic facies analysis, the work on thin sections [20], x-ray

Table 6.1 Summary of the regional geological setting that influenced the study areas through time including tectonics, climate, lithology, and sedimentary facies association

	Tectonic setting and environmental response	Palaeo—climate	Sedimentary facies; lithologies
Upper Rotliegend II	Saline perennial lake occupying central SPB, which underwent enlargement over time [21, short-lived marine incursions [47, 22]	Arid—semiarid [23]	Aeolian, sabkha, fluvial (wadi) and lacustrine [21, 24, 25, 47]
Zechsteln	Flooding caused by rifting in the Arctic North Atlantic contemporaneous to global sea level rise [49]; During the latest Permian a complex multidirectional rift system that transected the Variscan Fold Belt and the Central European Basin System (CEBS) [53, 54]	Arid; repeatedly restricted seawater influx [23]	Stacked evaporation cycles (marine clays, carbonates, calcium-sulphates and rock salts, potash and magnesium salt) [26, 27]
Buntsandsteln	Onset of tectonic activity postulated for the time since the Hardegsen tectonic phase [28, 29]; Several rifting pulses; NNW-SSE to NE-SW trending graben and half-graben systems [49]	Arid [23]	Red beds deposited under continental and shallow lacustrine, brakish to hyper-saline conditions [30, 31]
Muschelkalk	Tethys rifting through reactivated Variscan master faults accompanied by transmitted crustal motions from the Tethyan rift system to its northern periphery [32]; Tethys basin progressively compartmentalized into sub-basins [32]	Semiarid—arid [23]	Shallow restricted marine conditions to open marine Clearwater [49] Lower Muschelkalk: nodular and laminated lime muds [33, 34, 35]; Upper Muschelkalk: extensive carbonate platforms (Crinoidal limestones, laminated limestones to marls with Ceratites; [49])

(continued)

Table 6.1 (continued)

	Tectonic setting and environmental response	Palaeo—climate	Sedimentary facies; lithologies
Keuper	Enhanced differential subsidence, corresponding to increasing activity in the North Atlantic Rift System [36, 49]; Extensional tectonics in N–S striking zones (e.g. the GlUckstadt Graben), strike-slip movements of NW–SE striking structures (e.g. the Tornquist-Teisseyre lineament) (e.g. [37])	Semi-arid–humid [23]	Clastic fluvial to deltaic/estuarine system prograding gradually southwards from Fennoscandia, interfingering with marine, restricted-marine and partly evaporitic, brakish marine, and paralic environments along the SW basin margins [38, 39]
Jurassic	North Sea area affected by thermal uplift and erosion causing to massive clastic sedimentation [49] Arctic North Atlantic Rift System was active during entire Jurassic and only evolved into the break-up axis of Laurasia during the Late Cretaceous [49]	Semi-arid–tropical [23]	By the end of early Jurassic large parts of Western and Central Europe were occupied by epi-continental seas that provided a free communication between the Thetys and the Acrtic Sea [40] Marine sandstones of continental clastic shelf to marine clays and mudstones
Early Cretaceous	Crustal extension across the Arctic North Atlantic rift system, major tectono-eustatic sea-level variation and the overall increase in tectonic activity [49]; Late Cretaceous: new divergent plate boundaries in the Arctic North Atlantic and onset of convergence of Africa with southern margin of European craton [49]	Tropical [23]	Marine siliciclastics sediments deposited throughout the southernmost basins of the proto North Sea [49]; Late Cretaceous: Cenomanian to Danian chalk; later carbonate platforms [49]

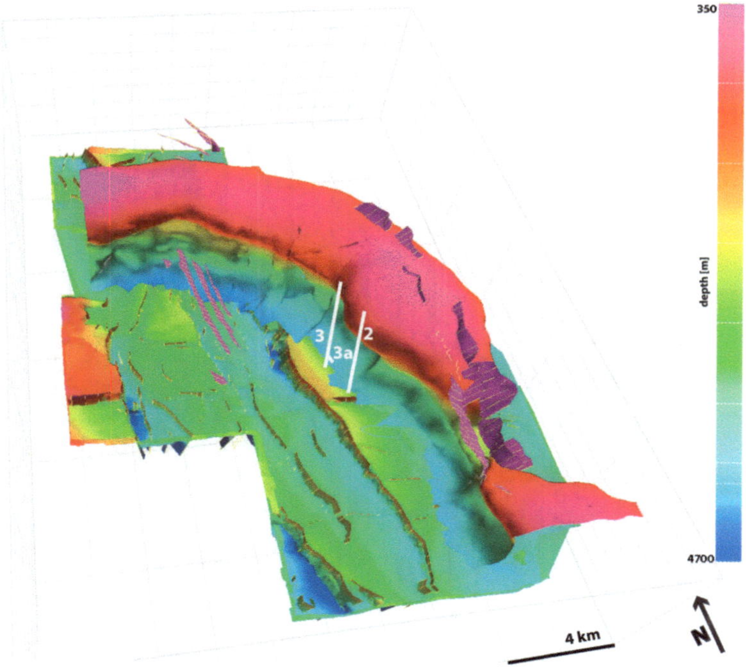

Fig. 6.2 Interpreted 3D seismic cube of the main study area with *top* Rotliegend depth map, faults and salt dome. Wells with cored Rotliegend strata are depicted in *white*

diffractometry [20], SEM imaging [41], cathodoluminescence and fluid inclusion measurements [18, 19] were taken into account. Previous reservoir evaluation studies, conducted by Lee [19], on fluid inclusions and the K/Ar dating of illite cements of the Upper Rotliegend II of Well 2 were taken into consideration when the burial and temperature history was derived by means of petroleum systems modelling.

In addition to the data set from the key study area, a 2D PSTM seismic line located in northern central Germany was used to verify the retro-deformation method (Fig. 6.4). The cross section was depth converted on the basis of a velocity model extracted from the well records of several wells. Well control is available from Well I, which is located in the central part of the 2D seismic line.

6.2.1 3D Isopach Data

Thirteen isopach maps of key stratigraphic units were calculated from interpreted seismic horizons (Fig. 6.5) involving Upper Rotliegend II, Zechstein, Lower Buntsandstein, Upper Buntsandstein (Röt Member), Lower Muschelkalk, Middle Muschelkalk, Upper Muschelkalk, Lower Keuper, Upper Keuper ± Jurassic,

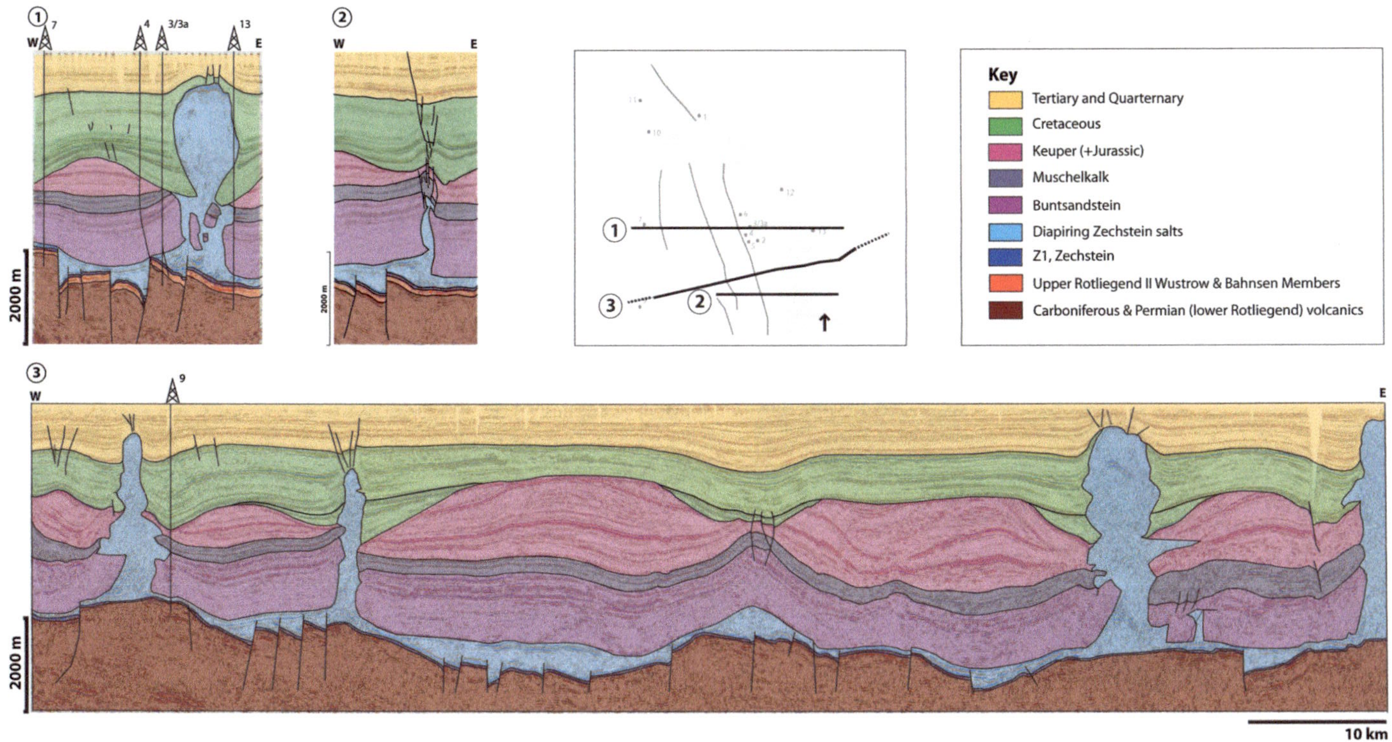

Fig. 6.3 Location of 2D Seismic Lines, well sites and major faults in the NW German study area (*outline grey-shaded*), which were used for retro-deformation (*yellow framed* in Fig. 6.1). Seismic interpretation and stratigraphic intervals depicted on original seismic *lines 1–3*

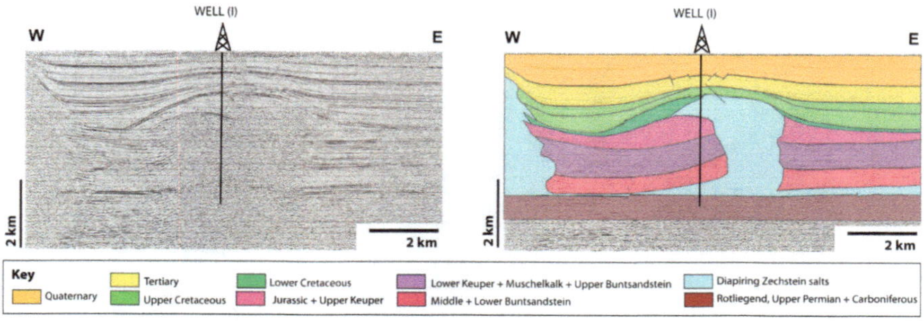

Fig. 6.4 *Left* Northern central German 2D seismic line used for retro-deformation (*orange framed* in Fig. 6.1). *Right* Seismic line with overlying interpretation and well position

Lower Cretaceous, Cenomanian—Santonian (early Upper Cretaceous), Campanian (Upper Cretaceous) and Maastrichtian (latest Upper Cretaceous). Differential sediment loads as response to regional and local tectonics and to salt kinematics provide an input parameter for the unravelling of the tectonic history. Areas of enhanced or reduced sediment thickness indicate tectonic movements, which in the study area, in most instances, are related to salt movement. Enhanced sediment thickness on the hangingwall of normal faults in contrast to reduced sediment thickness on the footwall suggests the existence of a faulting-induced palaeo-relief (Chap. 4).

6.2.2 2D Retro-Deformation

Ten interpreted stratigraphic horizons were used for a stepwise 2D retro-deformation (Figs. 6.6, 6.7, 6.8): Top Rotliegend, Top Zechstein, Top Solling (Upper Buntsandstein; as base of the Röt evaporites), Base Muschelkalk, Base Keuper, Base Upper Keuper, Base Cretaceous, Base Upper Cretaceous, Base Maastrichtian (latest Upper Cretaceous) and Base Tertiary (Fig. 6.3). The primary objective of the sequential restoration was to separate the various processes that have controlled the evolving basin geometry during specified time and to quantify their individual impact (e.g., Rowan [7]). As summarized by Rowan [7], several factors that control basin development have to be considered: (1) Sedimentation pattern: Variations in sediment thicknesses and facies, dependent on provenance and basin topography; (2) Compaction: The successive decrease in thickness and increase in density of sediment deposits during subsidence and burial; (3) Eustatic sea-level changes; (4) Fault-related deformation: In the case of the study area, dominantly represented by normal faulting and associated folding caused by regional extension and/or salt movement. Reverse faulting only rarely occurs in specific areas; (5) Salt Movement: Salt withdrawal, diapirism and lateral flow in response to sediment loading and regional extension; (6) Isostasy: The main effects are driven

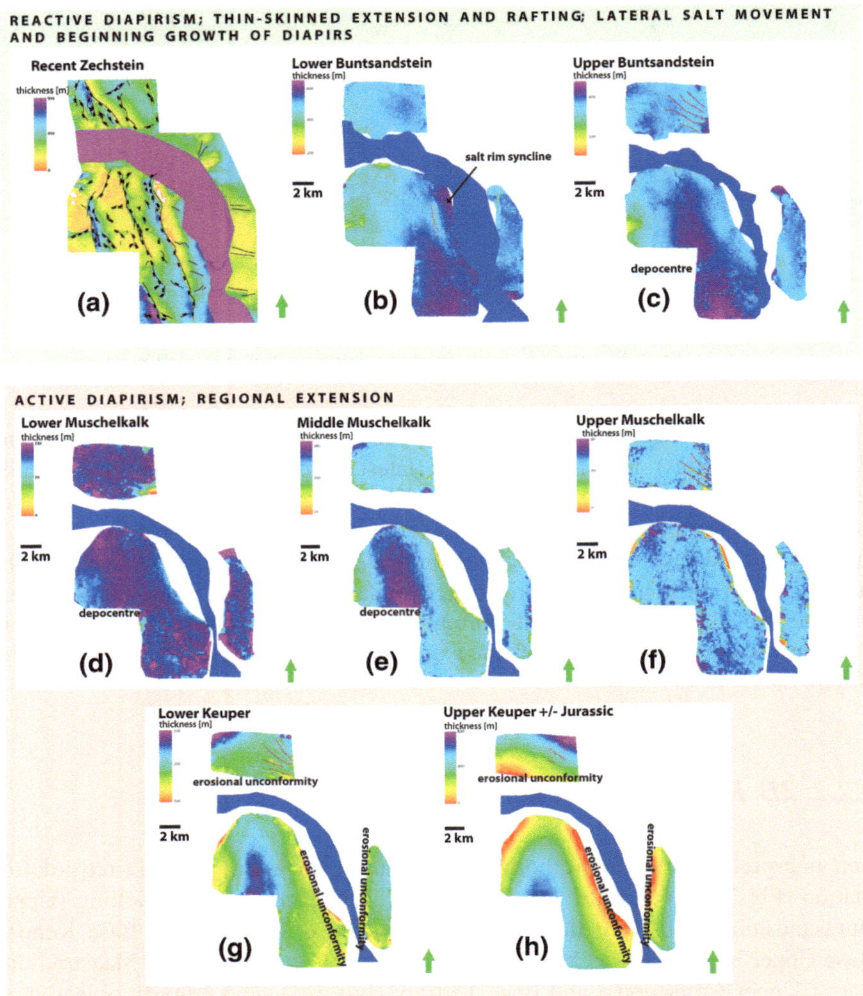

Fig. 6.5 Isopach maps of the study area from old to young. Plot is in context to different salt tectonic mechanisms

by subsidence due to sediment loading. It may include uplift or subsidence due to faulting and/or salt movement; (7) Thermal subsidence: Regional subsidence related to post-rift cooling.

Sedimentation pattern, salt movement, faulting, compaction, flexural isostasy, and thermal subsidence were considered as key input parameters for the retro-deformation of the study area. From young to old and after successive removal of strata, a decompaction with the parameters suitable for the particular lithology was applied to the cross sections. The applied decompaction algorithms are based on values published by Sclater and Christie [42] for sandstones, shales and

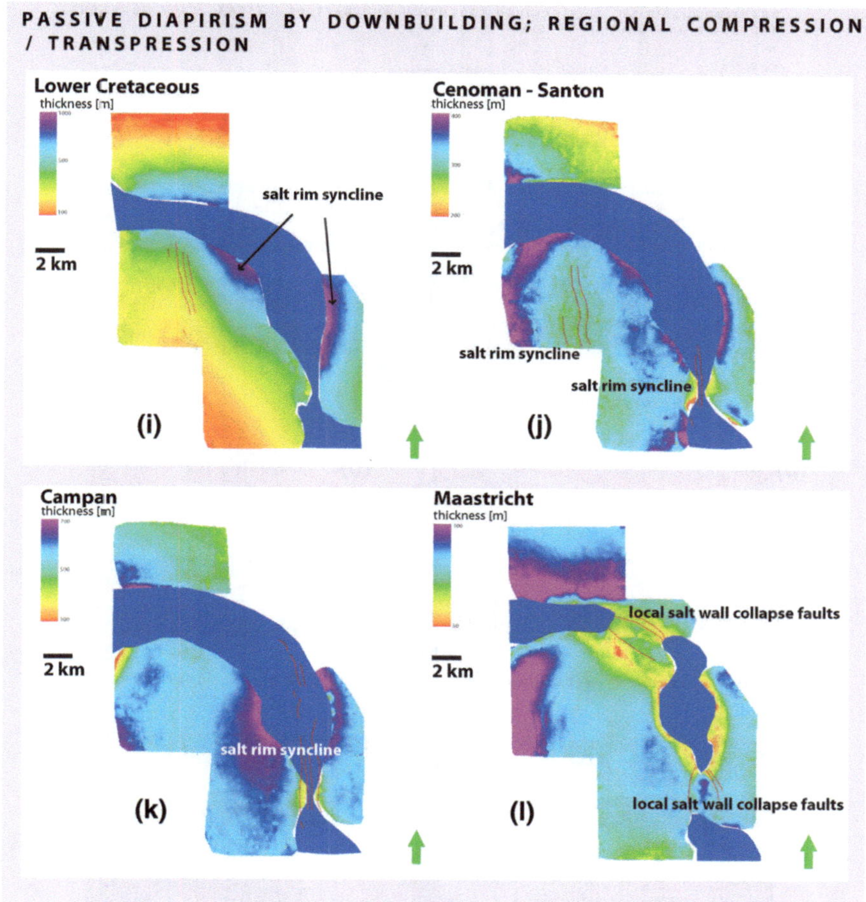

Fig. 6.5 continued

sandstone-shale mixtures, and by Schmoker and Halley [43] for limestones. Sediments were characterized from several well records and well logs. The "flexural isostasy" model was used [44], which turned out to only have a minor influence on section restoration. The basement is deflected by topographic and subsurface loads, and regional compensation occurs by plate flexure [44]. The relationship between topography and gravity anomaly due to the compensating mass at depth is length-scale dependent, and thus provides an estimate of the flexural strength of the elastic lithosphere within an analysis window [44]. However, the algorithms used in the retro-deformation were originally developed for brittle rock deformation and cannot easily reproduce the fluid-like viscous kinematics of salt [7]. Consequently, salt can only be passively restored and was therefore regarded as passive layer during the retro-deformation process. The salt area is estimated to be constant during time.

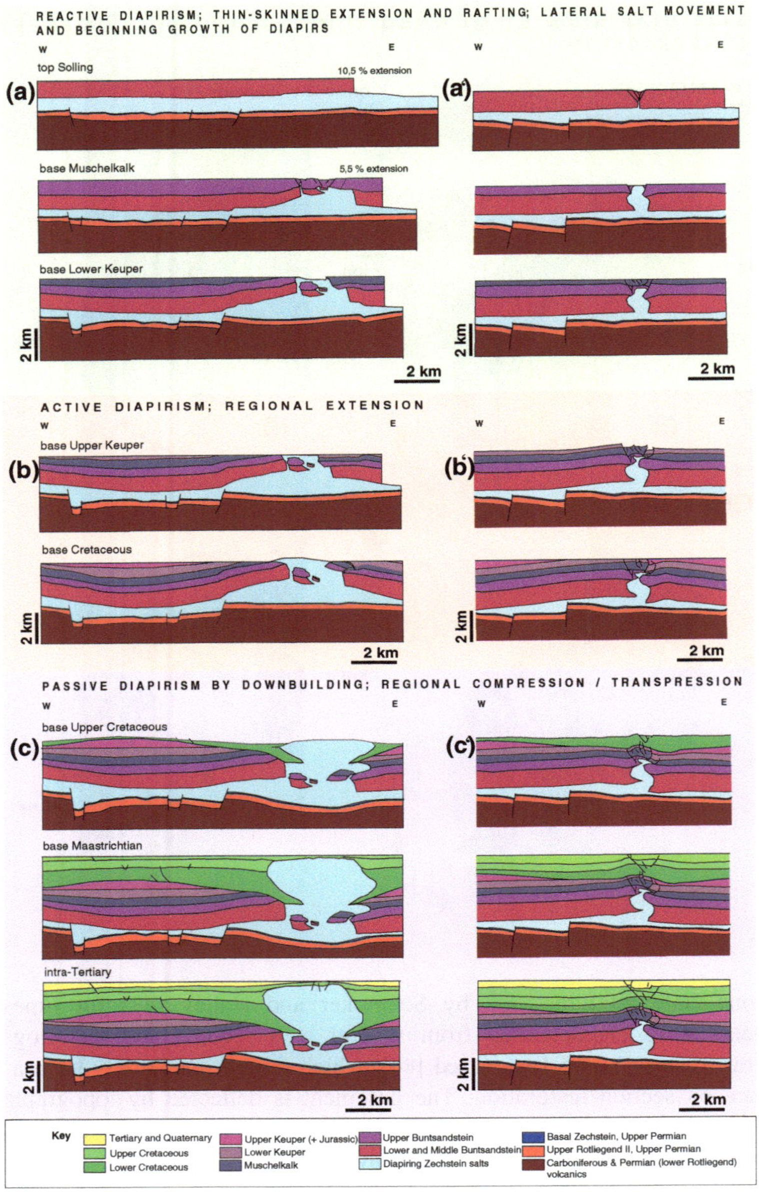

Fig. 6.6 Retro-deformed, depth-migrated seismic sections (see *lines 1* and *2* in Fig. 6.3), from the 3D seismic cube. Illustration of successive events in chronostratigraphic order, depicted in context of different salt tectonic mechanisms, indicated by coloured frames

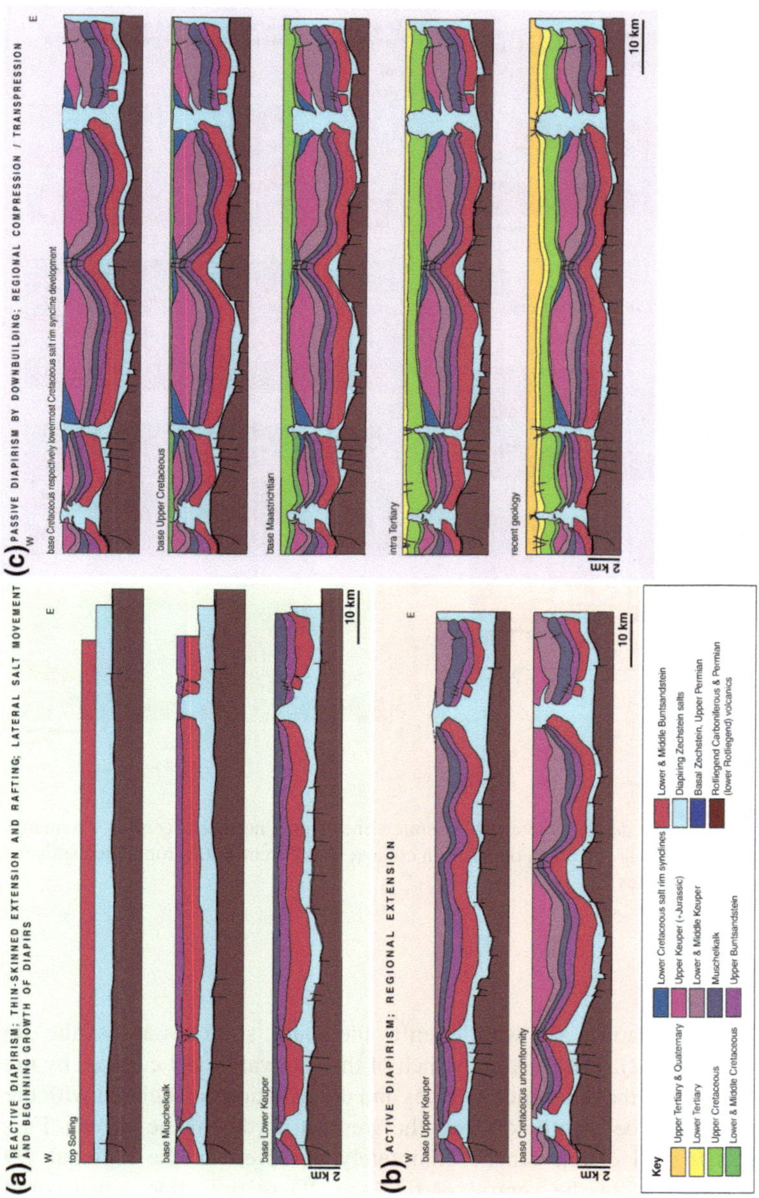

Fig. 6.7 Retro-deformed, depth converted regional 2D seismic line (see *line 3* in Fig. 6.3). Illustration of events from old to young (inverse), depicted in context of different salt tectonic mechanisms, indicated by coloured frames

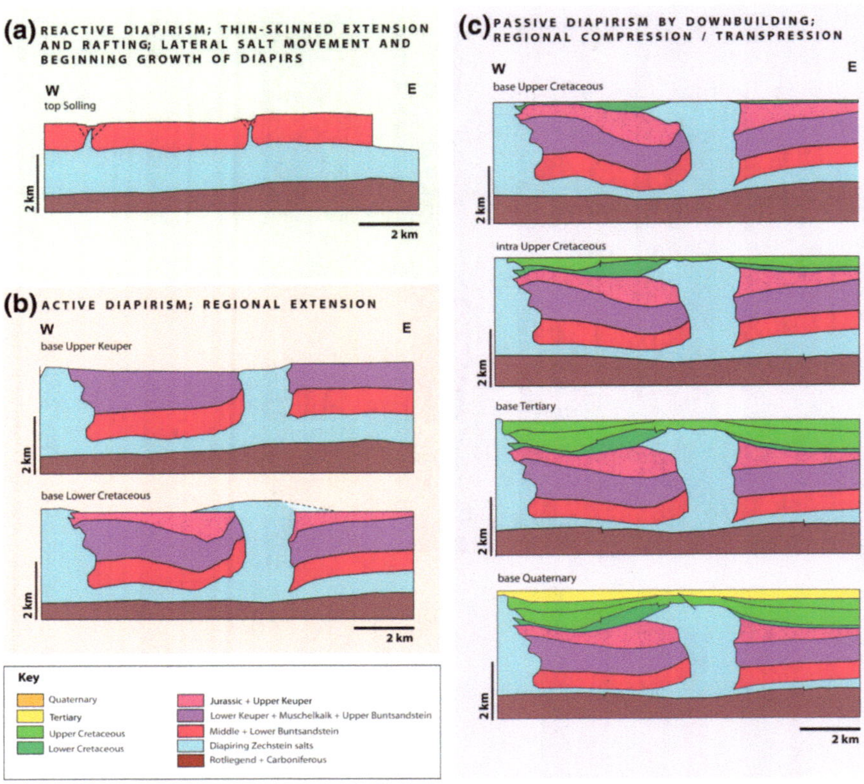

Fig. 6.8 Retro-deformed, depth converted seismic line from northern central Germany. Illustration from old to young (inverse), depicted in context of different salt tectonic mechanisms, indicated by coloured frames

6.3 Results

A U-shaped, mostly diapiric salt wall, open to the south, stretches across the 3D seismic survey (Fig. 6.2). The western branch of the salt wall is not covered by the 3D seismic survey, but the adjacent synclines and deformations associated with the salt accumulation can be identified along the western border of the survey. Furthermore, the salt wall extent can be completely observed on the regional 2D seismic line (Fig. 6.3). In the central part of the 2D seismic line a salt-cored anticline and in the eastern part a salt diapir with extensive overhangs are observed; the easternmost salt diapir bounds the seismic section to the E. The seismic section in northern central Germany shows two salt diapirs, one located at the centre of the line and one bordering the line to the W (Fig. 6.4).

Analogue modelling studies [12–14] suggest that salt cannot transmit large differential stresses due to its low viscosity. Therefore, post-salt series are assumed to be tectonically decoupled from the under burden by the Zechstein salt, acting as décollement surface (Figs. 6.3 and 6.4). In the following, the retro-deformation phases will be described from older to younger parts of the basin fill succession and pre- and post-salt times are described separately. The nearly E–W-oriented 2D sections were chosen for their characteristic deformation features. Both, the main basement and salt overburden faults are trending ~N–S. Due to the main E–W-deformed and N–S-elongated salt walls and diapirs, I assume that only N–S-oriented sections would show less to no salt kinematics through time.

6.3.1 3D Isopach Analysis and Lithologies

In the study area only late Upper Rotliegend II strata were deposited as response to tectonically induced marginal sedimentation in the SPB [45, 46]. The gas reservoir rocks comprise the fluvio-aeolian silt- to sandstones of the Wustrow and Bahnsen Members of the late Upper Rotliegend II [47]. Synsedimentary fault activity and existing palaeo-topography have been reconstructed from detailed isopach map interpretation by Vackiner et al. [47].

The total sediment thicknesses of the Zechstein vary between 50 and 750 m (Fig. 6.5a). Salt structures (domes, walls, and pillows) reach heights of up to 4,000 m. Top Zechstein is, apart from the accumulation in salt diapirs, located at 3,240 to 4,580 m depth. The Top Rotliegend seismic reflector is situated between 3,750 and 5,170 m depth. The thickness variations are mainly driven by salt withdrawal and extensive salt movement (Fig. 6.8a). Pre-existent fault zones and those that developed syn-tectonically to salt kinematics most likely acted as traps for residuals, like exotic breccias and anhydrite blocks, during salt buoyancy and entrainment into diapirs. Wells drilled in areas of depleted salt thickness mainly recovered Zechstein sediments (anhydrite, dolomite, limestone and claystone (especially the Kupferschiefer, which was deposited during the earliest Zechstein transgression) that remained in situ during salt kinematics. The isopach maps of the Zechstein 1 imply that Carboniferous to Rotliegend fault-induced topography was present during the deposition of the Lower Zechstein ([47]; Chap. 4). In contrast, the younger Zechstein (Z2 and younger) consistently covers and masks the initial structural grain.

The Lower and Middle Buntsandstein isopach maps (Fig. 6.5b) were calculated by subtracting Top Solling (2,580–3,930 m depth) from Top Zechstein (3,240–4,580 m depth). During the Lower Buntsandstein the local sedimentary environments were dominated by deposition of clays with some interbedded oolitic limestone and sandstone layers. Sediment thickness on the eastern side of the salt wall's western branch is 400–500 m. Along the western side of the salt wall's eastern branch Lower Buntsandstein thickness is up to 800 m (Fig. 6.5b). Middle Buntsandstein sections consist of sandstone and interbedded clay stones revealing slightly higher sediment thicknesses close to the salt wall, either from early development of salt rim synclines or the displacement of sediment by salt intrusion. For the Upper Buntsandstein isopach map (Fig. 6.5c) Top Solling (2,580–3,930 m depth) was used as the base boundary, and Top Buntsandstein (2,430–3,500 m depth) as the top boundary. While the Solling Member mainly consists of clay stones, the early Röt Member is dominated by halite and anhydrite. The late Röt Member comprises mainly pelites. A depocentre hosting sediments of up to 540 m thickness developed in the centre between the U-shaped salt wall branches. Sediments with a minimum thickness of 110 m in the western study area are located on the flanks of the salt wall. Keuper growth faults, which trend ENE–WSW and dip towards the SW, offset the Upper Buntsandstein strata in the northern study area in the SW (Fig. 6.3).

The Muschelkalk strata are bracketed by Top Buntsandstein ($\sim$2,430–3,500 m depth) and Base Keuper (2,460–3,230 m depth; Fig. 6.3). Isopach maps for all three subdivisions of the Muschelkalk (Lower; Fig. 6.5d, Middle; Fig. 6.5e, and Upper Muschelkalk; Fig. 6.5f) were generated. Lower Muschelkalk limestone, anhydrite and dolomite thicknesses are 70–210 m. The sedimentary depocentre between the branches of the U-shaped salt wall that evolved during the Upper Buntsandstein (Fig. 6.5d) was already active during the Middle and Lower Muschelkalk (Figs. 6.5e, f). During the Middle Muschelkalk marl, anhydrite and halite thicknesses were 80–190 m. In some places, the dolomite-to-limestone-dominated Upper Muschelkalk is truncated by a deeply incised Base Cretaceous unconformity. The Upper Muschelkalk sediment thicknesses range between 0 and 100 m. As observed from retro-deformed sections and isopach maps, the Upper Muschelkalk thickness is relatively constant with approximately 80 m.

The Jurassic and Keuper sediment sections are bounded by Top Muschelkalk (2,460–3,230 m depth) and the extensively folded, erosional unconformity of Base Cretaceous (1,840–3,090 m depth). The sedimentary succession of the Keuper is dominated by claystones. Lower Keuper thicknesses reach 0–700 m (Fig. 6.5g). This high range in thickness variation does not result from post-depositional Base Cretaceous incision, but also developed due to the formation of a syn-sedimentary depocentre, located between the branches of the salt wall. The syn-sedimentary tectonic regime was dominated by SW–NE extension with growth faults in the N oriented ENE–WSW and dipping towards SW (Fig. 6.5h). Sediment thicknesses of the Jurassic and Upper Keuper vary between 0 and 870 m (Fig. 6.5h). The lateral spread of the Jurassic and Upper Keuper is restricted by the erosional Base Cretaceous unconformity that cuts deeply into the Upper Keuper sediments. The presence of Jurassic strata can only be assumed, as they have not been drilled at the flanks of the salt wall. However, they are likely to be present in the centre of the study area, where sedimentary thickness maxima are located. Pronounced variability in sediment thicknesses is attributed to an active syn-sedimentary depocentre and to the erosional character of the Base Cretaceous unconformity towards the rims of the salt wall (Fig. 6.5h). ENE–WSW normal faults are located N of the salt wall and offset Upper Buntsandstein to Jurassic strata. They were active during the Jurassic and Upper Keuper. In addition, the activity of normal faults, displacing the area E of the salt wall against the sediments W of the salt wall, is confined to the Jurassic to Upper Keuper period. In general, sediment accumulation was higher in the eastern part of the study area (Fig. 6.5h).

The depth of the Base Cretaceous unconformity varies between 1,840 and 3,090 m. Top-Lower Cretaceous/base-Upper Cretaceous is located between 1,480 and 2,620 m depth. The Lower Cretaceous is strongly influenced by salt movement and the development of salt-related sedimentary depocentres (Fig. 6.5i). Sediment thicknesses vary between 30 and 1,270 m. Depocentres are indicated by sediment thicknesses of >1,000 m and directly fringe the diapiring salt wall (Fig. 6.5i). Since the Lower Cretaceous, the presence of extensive rim synclines can be determined along the flanks of the diapiring salt wall. 5 km west of the diapiric salt wall, in the centre of the study area, sediments are only ~300 m thick, and normal growth faults originate from flexural bending of the basin floor due to the enormous sediment load in the depocentres of the salt rim synclines (Fig. 6.5i). The faults were active during the Lower Cretaceous until the Santonian. Furthermore, some minor reverse faults in the Rotliegend, lower Zechstein, and the Carboniferous basement developed due to Cretaceous inversion tectonics along older, reactivated faults. As sediment thicknesses are high and salt deformation strongly varies the description of the Upper Cretaceous is subdivided into three parts: Cenomanian–Santonian (limestones and marls, Fig. 6.5j), Campanian (limestones and marls, Fig. 6.5k) and Maastrichtian (limestone, chalk, Fig. 6.5l). During Cenomanian—Santonian the sediment thicknesses vary between 50 and 740 m. The most extensive salt rim syncline, with the highest sediment thickness observed, developed along the western branch of the salt wall (Fig. 6.5j). High sediment thicknesses of up to 400 m are also observed in the rim syncline along the conjugated eastern salt wall branch. In contrast, the centre of the study area only exhibits a sediment thickness of ~280 m average. Extensional growth faults in the centre between the branches of the salt wall developed due to the high sediment load in the rim synclines since the Lower Cretaceous. Growth fault activity is indicated by enhanced hangingwall sediment thicknesses (Fig. 6.5j). The southern area of the eastern salt wall branch is narrowed. I assume that the salt accumulation along this constriction is only pillow shaped and salt residues and welds are left in places of former salt filling. Above the constriction, normal collapse-fault activity causes contrasting sediment thicknesses of 350 vs. 750 m in footwall and hangingwall block settings (Fig. 6.5j). During the Campanian, salt rim synclines hosting sediments with up to 700 m thickness were active and indicate that salt has been located close to the surface. In contrast, only up to 450 m of sediments were accumulated in the centre of the study area. The salt wall further collapsed on top of the constricted area (Fig. 6.5k). The thickness of the Maastrichtian is 30–400 m. Ongoing development of salt rim synclines with sediment thicknesses of up to 400 m along the eastern side of the western salt wall branch and fringing the salt wall in the north document the evidence for halokinesis (Fig. 6.5l). Above the constriction normal collapse-faults border areas of enhanced hangingwall sediment thickness. Most of the faults on top of the salt accumulation developed during Maastrichtian times. An additional smaller collapse along the north-eastern bending of the salt wall is only affecting the Maastrichtian sediments. It is bordered by several normal faults that are dipping towards the central salt diapir. The slightly enhanced sediment thicknesses in these areas cannot be ascribed to a salt collapse (in the size of the structure in the constriction), but rather to lateral salt movement, e.g. towards the northern and western part of the salt wall, where extensive salt rim synclines have been mapped (Fig. 6.5l).

The Tertiary and Quaternary sedimentation pattern are affected by collapse faulting that borders areas of increased sediment thickness on top of the salt wall. Both, fault activity and salt movement is still ongoing today. The Tertiary mainly consists of claystones, whereas Quaternary and latest Tertiary sections are dominated by coarse grained sands and fine grained gravels.

6.3.2 2D Retro-Deformation

The Upper Rotliegend II strata underwent multiple deformation phases, which reactivated Carboniferous and Permian structures, associated with multi-diagenetic overprint of fluvio-aeolian sandstones [1, 48, 49]. In the following, the deformation phases are assigned to salt tectonic movements. For the lowermost Zechstein synsedimentary tectonics can be reconstructed. In contrast, the upper Zechstein was not affected by synsedimentary deformation. Brittle intra-salt fragments that consist of Buntsandstein to Muschelkalk lithologies and are entirely encased in salt were mapped in the 3D seismic data (Figs. 6.3, 6.6a–c). Kinematic modelling provides no further constraints on whether the fragments sunk or ascended into the salt.

The retro-deformation (Fig. 6.6a‘–c‘) indicates an early collapse phase in a restricted area above the salt wall, bordered by active normal collapse-faults, during the Muschelkalk (Fig. 6.6a‘). Because of this fact, the Muschelkalk thickness on the hangingwall of these faults is higher than on the footwall.

For the easternmost salt dome in the 2D seismic line (Figs. 6.3, 6.7 and 6.9a) a huge potential salt namakier, directly overlying Top Lower Keuper, was mapped (Fig. 6.7b). Its lateral extension is ~5 km and its thickness reaches up to 200 m close to the salt diapir stem. It cuts erosionally into the underlying sediments (Fig. 6.9a), indicating subaerial exposure. Against the top sedimentary onlaps respectively downlaps are observed (Fig. 6.9a). A salt intrusion would have conformable contacts to the over- and underlying sediments. However, several additional, but smaller salt namakiers or intrusions of up to 10,000 m^2 are observed in the Lower Keuper interval. Jurassic salt rim synclines are observed in the 2D regional seismic line (Fig. 6.7b). A salt pillow in the centre of the 2D seismic line (Fig. 6.3), started downbuilding since the Upper Triassic, reached its most pronounced elevation during the Upper Keuper (Fig. 6.7b). Another huge potential salt namakier with lateral extensions of 1 km and a thickness of up to 900 m, which discordantly overlies the Base Cretaceous, was mapped in northern central Germany (Figs. 6.4, 6.8 and 6.9b). Again, the underburden sediments are erosionally truncated by the salt and the overburden sediments downlap or onlap against it (Figs. 6.8b, 6.9b).

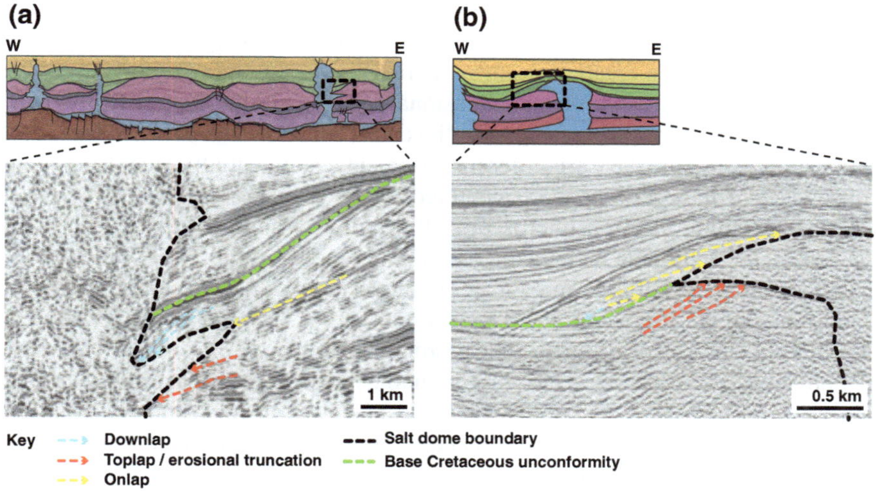

Fig. 6.9 Zoom of the areas of huge potential salt namakiers with lap geometries of underlying and overlying sediments displayed. *Left* salt namakier in 2D Regional seismic line in north-western Germany. *Right* Salt namakier in northern central German 2D seismic line

For the Lower Cretaceous, the activity of normal growth faults, which originate from bending of the sedimentary base due to the enormous sediment load in the depocentres of the salt rim synclines, can be assigned to the Lower Cretaceous to the Santonian deposition (Fig. 6.6c). The development of Upper Cretaceous to Tertiary collapse faults on top of the salt wall is reconstructed in detail (Fig. 6.6a'–c'). In contrast to the Cretaceous synsedimentary tectonic setting, the Tertiary is rather unaffected by tectonic deformation (Figs. 6.6, 6.7, 6.8).

6.3.3 Core Data (Sandstone Petrography, Fluid Inclusions)

Core data are only available from the main study area in north-western Germany. Wells used for additional analysis of sandstone petrography are located in the footwall of a major fault ~0.6 km (Well 3) and ~0.9 km (Well 3a) away from the fault zone (Fig. 6.2). Microthermometry [18, 19] reveals several stages of fluid inclusion formation. An assumed geothermal gradient of 33 °C/km depth has been confirmed by temperature curves from well logs. Fluid inclusions in carbonate and quartz cements of the Wells 2 and 3 mainly consist of water, sodium chlorite and calcium chlorite, revealing temperatures of ~140–170 °C. Microthermometry of quartz cements from quartz and calcite cements of Well 3a, at larger distance to the fault zone, are not as high (~120–150 °C). Formation of cement minerals with fluid inclusion formation temperatures of ~145–170 °C are assigned to the burial depth of the Upper Rotliegend II during the early Upper Cretaceous; temperatures of up to ~120–145 °C are attributed to the Lower Cretaceous to Upper Jurassic burial depth. Fluid inclusion studies and K–Ar dating in the study area, conducted by Lee [19], conclude that the phases of illite and quartz precipitations overlapped and that the main cementation occurred during the Upper Jurassic to Lower Cretaceous doming event and related extensional tectonics in the central North Sea.

Within the reservoir rocks (Upper Rotliegend II Wustrow and Bahnsen Member), which directly overlie the Carboniferous basement and potentially older Rotliegend volcanics, five diagenetic steps are identified by petrographic, cathodoluminescence [20] and scanning electron microscopy (SEM) image analyses ([41]; Fig. 6.10). These data are combined with the thermal history model in Fig. 6.11, which shows that the main phases of deformation are the Jurassic, Lower Cretaceous, and Upper Cretaceous. In the following, the diagenetic phases as reported by Havenith et al. [20], will be described in chronological order (Fig. 6.10).

Eodiagenesis: Illite grew as discontinuous tangential rims around detrital mineral grains. Complete coatings can only be found in rock units close to the underlying volcanics and volcanic breccias (Fig. 6.10a3). This indicates a synsedimentary admixture of altered volcanic material to the quartz-dominated aeolian sands.

Mesodiagenesis: (1) Under reducing conditions early, Fe-rich chlorite grew as a rim of radial fibres enveloping pore spaces. Chemical components like Al^{3+} and Si^{4+} might have been mobilized by the onset of compaction. (2) Fe-dolomite segregated into dolomite and ankerite which, therefore, often occur in textural equilibrium (Fig. 6.10a4). (3) Further mineral overgrowth and pressure solution of quartz and the neo-formation of platy illite and late chlorite essentially reduced the intergranular pore space (Fig. 6.10b4). Chlorite incorporates Fe^{2+} and Mg^{2+}, which were released during reaction of dolomite and ankerite to calcite. Albitic feldspars are observed as overgrowths, forming euhedral crystals, which grew into open pore spaces (Fig. 6.10b2). Fibrous and meshwork-type illite essentially cement the pore space. The presence of authigenic rutile is interpreted to result from hydrocarbon migration [50]. (4) Ongoing quartz overgrowth (Fig. 6.10c2) and fibrous illite and chlorite precipitation (Fig. 6.10c1) in pore space were related to a late phase mineralization. Chemical components necessary for the formation of illite were most probably derived from the dissolution of feldspar and early illite coats and plates (Fig. 6.10a1, a2). (5) The last diagenetic episode observed includes late quartz overgrowths, as well as the precipitation of anhydrite, calcite, and barite, potentially originating from Zechstein brines (Fig. 6.10c3, c4).

6.4 Interpretation

In this study, data from isopach maps (Fig. 6.5) has been analyzed and a retro-deformation of interpreted seismic sections (Figs. 6.6, 6.7, 6.8) to define different salt rise mechanisms and assign them to particular tectonic phases and their diagenetic effects (Fig. 6.10) on the Upper Rotliegend II tight reservoir rocks has been undertaken (Fig. 6.11). The U-shaped salt wall follows the salt rise

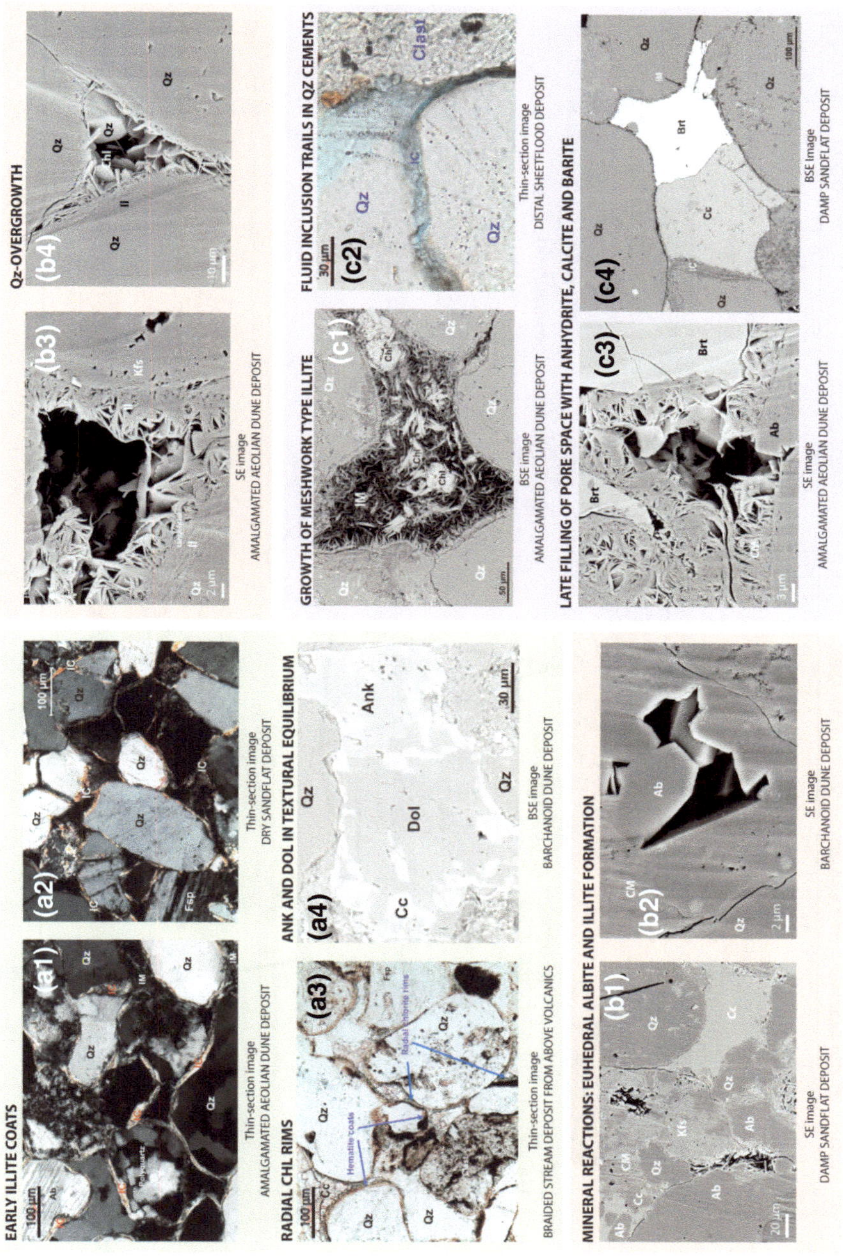

Fig. 6.10 Thin-section, BSE and SEM images (a1 to c4) illustrating the diagenetic history through time, from old to young. Images are grouped in the context of the different deformation mechanisms. Mineral abbrevaitions: *Qz* = quartz, *Ab* = albite, *Kfs* = potassium feldspar, *Fsp* = feldspar, undifferentiated, *Cc* = calcite, *Brt* = barite, *Chl* = chorite, *Il* = illite, *IM* = meshwork illite, *IC* = cutaneous illite rims, *CM* = clay minerals, undifferentiated

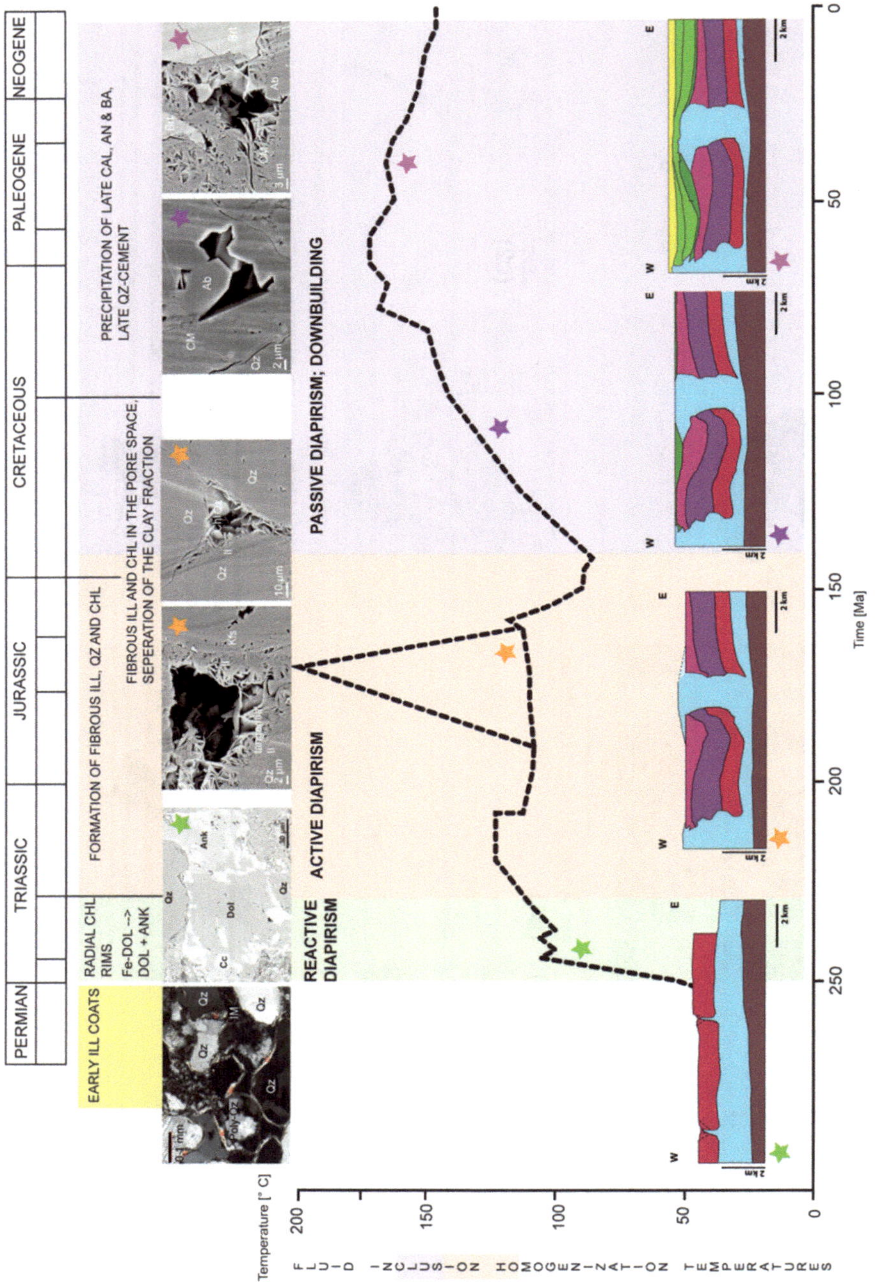
PERMIAN
TRIASSIC
JURASSIC
CRETACEOUS
PALEOGENE
NEOGENE
EARLY ILL COATS
RADIAL CHL RIMS
Fe-DOL --> DOL + ANK
FORMATION OF FIBROUS ILL, QZ AND CHL
FIBROUS ILL AND CHL IN THE PORE SPACE, SEPERATION OF THE CLAY FRACTION
PRECIPITATION OF LATE CAL, AN & BA, LATE QZ-CEMENT
REACTIVE DIAPIRISM
ACTIVE DIAPIRISM
PASSIVE DIAPIRISM; DOWNBUILDING
W
E
2 km
Temperature [° C]
FLUID INCLUSION HOMOGENIZATION TEMPERATURES
200
150
100
50
0
250
Time [Ma]

◀**Fig. 6.11** Reservoir temperature history (*dashed line*) extracted from the petroleum systems model with (*upper dashed line*) and without (*lower dashed line*) hydrothermal peak during the Jurassic (Blumenstein-Weingartz 2010) in relation to the different phases of salt tectonic mechanisms and diagenesis. Fluid inclusion homogenization temperatures ([18], [19]) at the temperature curve correspond to the diagenetic mechanisms and the salt tectonics as indicated by the same colour of the stars. Mineral abbrevaitions: *Qz* = quartz, *Ab* = albite, *Kfs* = potassium feldspar, *Fsp* = feldspar, undifferentiated, *Cc* = calcite, *Brt* = barite, *Chl* = chorite, *Il* = illite, *IM* = meshwork illite, *IC* = cutaneous illite rims, *CM* = clay minerals, undifferentiated

mechanisms, reactive, active and passive diapirism, defined by Vendeville and Jackson [10]. Its development alters, probably due to changes in the rates of tectonic extension, depletion of the source layer or changes in sedimentation rates ([51]; Figs. 6.6, 6.7, 6.8). 2D retro-deformation uncertainties comprise the 3D spatial variation of salt deformation, salt out-of section migration and mass quantification of eroded salt during subaerial exposure (Figs. 6.7b, 6.8b). Consequently, the detailed quantification of a salt area in 2D is therefore not possible. I assume that the salt area is constant during time and regard the sediment thicknesses as proxy for the definition of salt tectonic mechanisms. In the following, the most likely salt rise mechanisms will be described stepwise in chronological order.

To the top of the Zechstein retro-deformed sections reveal that the original thickness of the Zechstein was ~750–900 m. Within the retro-deformed sections in the central part of the study area the Zechstein thickness was slightly higher (Fig. 6.6a). The fault-controlled Carboniferous to Rotliegend palaeo-topography was onlapped and finally covered by the Zechstein transgression. As top Zechstein was most likely developed as a horizontal surface prior to the onset of salt movements, variances in Zechstein thickness are attributed to the underburden topography. Early tectonic events during the Zechstein are described by Geluk [52, 53] from the Netherlands and by Legler et al. [46] from Northern Germany. Resulting thickness variations from foot- to hangingwall blocks are obvious in isopach maps of the Zechstein Z1 ([47]; Chap. 4). An early Zechstein tectonic event, contemporaneous to the Tubantian I tectonic event and described by Geluk [52], is assumed to have influenced the study area as well. In contrast, the late Zechstein is postulated to have been a tectonically relatively quiet period. Although there is only evidence of minor syn-depositional Zechstein faulting in the study area, basin subsidence cannot be solely explained by the effects of cooling and compaction following an Early Permian thermal anomaly [48].

Contemporaneous to post-sedimentary in situ alteration, the Upper Rotliegend II sandstones developed discontinuous rims of illite around detrital quartz and feldspar grains (Fig. 6.10a3). Similar observations concerning early diagenesis were e.g. described by Schöner and Gaupp [54].

During the Lower Buntsandstein lateral salt movement, salt withdrawal and probably breaching of salt through the overburden initiated as a consequence of tectonically induced differential loading in the overburden (Figs. 6.6a, a‘, 6.7a, 6.8a). The rate of salt rise during reactive diapirism is entirely determined by the rate of extension. A modern analogue example for this diapirism phase is given by viscous salt flow beneath the brittle overburden in Canyonlands National Park in

Utah, Unites States of America [55]. Lateral salt movement accommodates deformation that results in a series of regular spaced grabens. Reactive diapirs developed preferentially at localities characterized by enhanced graben displacement [55]. The retro-deformation results imply minor thin-skinned extension that is most likely the cause for the initial phase of salt movement. After Mohr et al. [56] and Kukla et al. [57] the major phase of thin-skinned extension took place during the Upper Buntsandstein, leading to detached faulting in the sedimentary cover of the Buntsandstein. When syn-depositional faults opened and cut near the emergent diapir crest, salt extrusion and diapirism began [58]. During further extension the small, initial salt walls widened (Figs. 6.6a, a‘, 6.7a, 6.8a). Salt piercement that did not evolve to diapirs remained as salt welds, thin layers of salt, or brecciated, insoluble residue, remaining in the overburden strata ([59]; Fig. 6.7). Early and middle Triassic salt tectonics are frequently caused by reactive diapirism as a reaction to thin-skinned extension (Figs. 6.6a, a‘, 6.7a, 6.8a). During the latest Muschelkalk, the diapiring salt wall was at least situated directly below the sediment surface or, most likely, reached the surface. The diagenetic response to the tectonic overprinting is the breakdown of Fe-dolomite to form dolomite and ankerite that can be found to occur in textural equilibrium (Fig. 6.10a4).

Most of the Jurassic strata and also parts of the Keuper were eroded during the early Cretaceous inversion tectonics (Figs. 6.6c, c‘, 6.7c, 6.8c). Two huge salt namakiers are mapped at Keuper and Jurassic levels (Fig. 6.9), and are indicators for active diapirism during regional extension. Salt namakiers differ from salt sheets by their geometry. They erosionally truncate the underlying strata while salt intrusions only inflate the strata without truncation ([51]; Fig. 6.9). In the study areas, the Lower Keuper strata (NW Germany), respectively Upper Keuper and/or Jurassic strata (northern central Germany), top-lap against erosional unconformities that are overlain by salt namakier deposits (Fig. 6.9). The overburden sediments (Lower Cretaceous or Upper Keuper) show toplap or downlap against the salt. The active intrusion of salt diapirs always causes a space problem in the overburden and is only possible if overlying sediments are relatively thin, whereas the salt pressure has to exceed the brittle strength of the overburden ([51]; Figs. 6.7b, 6.8b). Once active diapirism was initiated, the salt diapir pierces each new sediment increment that attempts to bury it [51]. Salt namakiers, in this case, are only interpreted on one side of the diapir stem in the seismic section. In the retro-deformed sections equal namakiers are postulated to be located on the opposite side of the concerned diapir stem. If the namakier was initially developed symmetrically, one of the extrusive salt layers was eroded or leached during subaerial exposure. In addition to salt glaciers, monoclines with oval bended faults developed and huge Buntsandstein and Upper Buntsandstein—Muschelkalk brittle intra-salt fragments (up to $\sim 1\ km^3$) were moved inside the salt section (Fig. 6.6b). The difference in timing of salt glacier development in northern Germany compared to the study area suggests that subaerial exposure of salt varied in age. At the south-western SPB margin, subaerial salt exposure is observed as early as during Lower Keuper, whereas at the southern margin of the SPB, it occurred only during the Upper Jurassic/Lower Cretaceous. In consequence, it is assumed, that the geodynamic characteristics of active diapirism were also lasting longer at the SPB's southern margin in comparison to the SPB's south-western margin. Temperatures of ~ 120–145 °C match the Jurassic (to Lower Cretaceous) burial depth. Intergranular cements developed contemporaneous to the change of salt kinematics from active to passive diapirism. Mineral reactions of the mesodiagenetic phase (iii) include the neo-formation of calcite cements, quartz overgrowth (Fig. 6.10b4), pressure solution and the precipitation of late chlorite (Fig. 6.10b3). K/Ar-ages of authigenic illite also indicate Jurassic ages [19] and confirm the microthermometry results.

A phase of passive diapirism and large-scale sediment downbuilding started during the (Jurassic/) Lower Cretaceous (Figs. 6.6c, c', 6.7c, 6.8c), and is still ongoing. Sediment downbuilding causes growing of the salt diapir contemporaneously to sinking of the base of the salt structure, while the crest of the diapir remains at the surface. During the downbuilding phase, extensive salt rim synclines develop (Figs. 6.6c, 6.7c). Due to the locally high sedimentary load in the salt rim synclines, growth faults established distal to the salt structure in the centre of the study area (Figs. 6.5i, j, 6.6c). The Late Cretaceous to Tertiary is then influenced by the Alpine compressive deformation and the opening of the Northern Atlantic [48, 60]. Local inversion structures can be observed cutting into the salt caprock and the underburden (e.g. Fig. 6.5h). They are associated with salt being compressed in the diapir stem and withdrawn into the diapir. The fluid inclusion homogenization temperatures of ~ 145–170 °C can be assigned to an Upper Cretaceous burial depth and stress regime [mesodiagenetic phase (v)], during which further growth of meshwork type illite and late chlorite (Fig. 6.10c1), further quartz cementation (Fig. 6.10c2), calcite pressure solution and overgrowth occurred. The Upper Cretaceous downbuilding is contemporaneous to the late diagenetic filling of pores by anhydrite, calcite and barite (Fig. 6.10c3, 4).

6.5 Conclusions

Since the study area is strongly influenced by the Zechstein salt, I was able to assign major phases of halokinesis to regional tectonic events. Early or late regional extension and shortening caused three successive stages of salt-diapir growth and contemporaneous depocenter development. The structural and stratigraphic evolution of the study areas involved the rise of salt structures from the late Permian until today. Salt structures are of variable volume and architecture, including simple salt-cored anticlines as well as complex salt diapirs with several extensive overhangs, which are most likely fossil salt glaciers, embedded into the stratigraphic record adjacent to the salt structure. Different salt rise related depocentres developed over time: Salt rim synclines and collapse respectively withdrawal related growth faults on top of the salt wall host enlarged Buntsandstein to Tertiary sediment thicknesses of up to 700 m.

1. Salt movements started with lateral salt flow. Salt diapirism and rafting were initiated by tectonically induced differential loading during thin-skinned extension. Extensional tectonics caused small salt injections into the overburden. Some injections grew to diapirs during a raft tectonic phase in the Upper Buntsandstein and Muschelkalk. Mineral reactions during this phase did not degrade the primary compositionally and texturally mature gas reservoir sandstones.
2. Keuper to Jurassic processes were dominated by active salt diapirism, triggered by regional extension with a maximum extension rate during Jurassic. The Jurassic peak temperatures, which accompanied the change from active to passive diapirism, triggered quartz overgrowth and the formation of authigenic clay minerals [20]. The salt pressure exceeded the brittle strength of the overburden during active diapirism. The geometry of the diapiring salt wall with its Keuper and Jurassic salt namakiers is shaped by several periods of subaerial exposure of salt, occurring when the rate of upwelling was higher than the rate of dissolution, aggradation and extension.
3. Since the Lower Cretaceous, the diapiring salt wall was rising due to passive diapirism and extensive salt rim syncline development. This phase was influenced by late-stage pore filling of sulphates and carbonates [20].

Contemporaneous to salt movement, or even influenced by it, the Upper Rotliegend II reservoir rock underwent several diagenetic phases, which led to a reduction of the sandstone reservoir quality through time. Linking the diagenetic phases to the regional structural development clearly shows, that the main phases of quartz overgrowth, pressure solution in quartz and calcite, and the growth of pore and pore throat cementing clay minerals occurred contemporaneously to the main stages of salt-diapir development induced by regional tectonics (Fig. 6.11).

References

1. Lohr T, Krawczyk M, Tanner DC, Samiee R, Endres H, Oncken O, Kukla PA (2007) Strain partitioning due to salt; insights from interpretation of a 3D seismic data seit in the NW German Basin. Basin Res 19(4):579–597
2. Stollhofen H, Bachmann NGH, Barnasch J, Bayer U, Beutler G, Franz M, Kästner M, Legler B, Mutterlose J, Radies D (2008) Upper Rotliegend to early Cretaceous Basin development. In: Littke R, Bayer U, Gajewski D, Nelskamp S (eds) Dynamics of complex intracontinental Basins; the Central European Basin system. Springer, Berlin, pp 181–210
3. Roberts AM, Kusznir NJ, Yielding G, Styles P (1998) 2D flexural backstripping of extensional basins; the need for a sideways glance. Pet Geosci 4:327–338
4. Dahlstrom CDA (1969) Balanced cross sections. Can J Earth Sci=Revue Canadienne des Sciences de la Terre 6(4, Part 1):743–757
5. Gibbs AD (1983) Balanced cross-section construction from seismic sections in areas of extensional tectonics. J Struct Geol 5(2):153–160 (Balanced cross-sections and their geological significance; a memorial to David Elliott)
6. Hossack JR, McGuinness DP (1990) Balanced sections and the development of fault and salt structures in the Gulf of Mexico (GOM). In: Proceedings of Geological Society of America

1990 annual meeting. Abstracts with programs—Geological Society of America, vol 22(7). p. 48
7. Rowan MG (1993) A systematic technique for the sequential restoration of salt structures. Tectonophysics 228(3–4):331–348 (New insights into salt tectonics; collection of invited papers reflecting the recent developments in the field of salt tectonics, Cobbold)
8. Bishop DJ, Buchanan PG, Bishop CJ (1995) Gravity-driven thin-skinned extension above Zechstein group evaporites in the western central North Sea; an application of computer-aided section restoration techniques. Mar Pet Geol 12(2):115–135
9. Buchanan PG, Bishop DJ, Hood DN (1996) Development of salt-related structures in the central North Sea; results from section balancing. In: Alsop GI, Blundell DJ, Davison I (eds) Salt tectonics, vol 100. Geological Society Special Publications, London, pp 111–128
10. Vendeville BC, Jackson MPA (1992) The rise of diapirs during thin-skinned extension. In: Jackson MPA (ed) Special issue; salt tectonics, vol 9(4). Marine and Petroleum Geology, Angola, pp 331–353
11. Schultz-Ela DD, Jackson MPA, Vendeville BC (1993) Mechanics of active salt diapirism. Tectonophysics 228(3–4):275–312 (New insights into salt tectonics; collection of invited papers reflecting the recent developments in the field of salt tectonics)
12. Koyi H, Talbot CJ, Torudbakken BO (1993) Salt diapirs of the Southwest Nordkapp Basin; analogue modelling. Tectonophysics 228(3–4):167–187 (New insights into salt tectonics; collection of invited papers reflecting the recent developments in the field of salt tectonics)
13. Nalpas T, Brun JP (1993) Salt flow and diapirism related to extension at crustal scale. Tectonophysics 228(3–4):349–362 (New insights into salt tectonics; collection of invited papers reflecting the recent developments in the field of salt tectonics)
14. Vendeville BC, Ge H, Jackson MPA (1995) Scale models of salt tectonics during basement-involved extension. Pet Geosci 1:179–183
15. van Keken PE, Spiers CJ, van den Berg AP, Muyzert EJ (1993) The effective viscosity of rocksalt; implementation of steady-state creep laws in numerical models of salt diapirism. Tectonophysics 225(4):457–476
16. Poliakov ANB, Podladchikov Y, Talbot C (1993) Initiation of salt diapirs with frictional overburdens; numerical experiments. Tectonophysics 228(3–4):199–210 (New insights into salt tectonics; collection of invited papers reflecting the recent developments in the field of salt tectonics)
17. Podladchikov Y, Talbot C, Poliakov ANB (1993) Numerical models of complex diapirs. Tectonophysics 228(3–4):189–198 (New insights into salt tectonics; collection of invited papers reflecting the recent developments in the field of salt tectonics)
18. Havenith VMJ (in prep.) Diagenese evolution von Oberrotliegend Sandsteinen in Ostfriesland. Dissertation
19. Lee M (1996) Diagenesis of the Rotliegend sandstone of southern Ostfriesland. Structural and stratigraphic processes MEPTEC. Confidential report, Dallas, Texas, USA, p. 30
20. Havenith VMJ, Meyer FM, Sindern S (2010) Diagenetic evolution of a tight gas field in NW Germany. In: Proceedings of DGMK/ÖGEW-Frühjahrstagung 2010, Fachbereich Aufsuchung und Gewinnung, Celle
21 Glennie KW (1990) Introduction to the petroleum geology of the North sea 3rd ed. Blackwell Scientific, Oxford
22 Stollhofen H, Bachman GH, Barnasch J, Bayer U, Beutler G, Franz M, Kästner M, Legler B, Mutterlose J, Radies D (2008) Upper rotliegend to early cretaceous basin development. In: Littke R, Bayer U, Gajewski D, Nelskamp S (Eds) Dynamics of complex intracontinental basins. The central european basin system, pp 181–210
23 Scotese C. (2008) Palaeomap project
24 George GT, Berry JK (1993) A new palaeogeographic and depositional model for the upper rotliegend of the UK sector of the Southern North Sea. In: North CP, Prosser DJ (eds) Characterization of fluvial and aeolian reservoirs, Special Publication, Geological Society of London, vol 73, pp 291–319

25 Strömbäck AC, Howell JA (2002) Predicting distribution of remobilized aeolian facies using sub-surface data: the Weissliegend of the UK Southern North Sea. Petroleum Geoscience 8:237–249
26 Peryt TM, Wagner R (1998) Zechstein evaporite deposition in the Central European Basin: cycles and stratigraphic sequences. J Seismic Explor 7(3-4):201–218
27 Warren JK (2006) Evaporites: sediments, resources and hydrocarbons. Heidelberg, Springer p 1036
28 Frisch U, Kockel F (1997) Altkimmerische bewegungen in nordwestdeutschland. Brandenburger Geowiss Beitr 4(1):19–29
29 Brückner-Röhling S, Röhling HG (1998) Palaeotectonics in the Lower and Middle Triassic (Buntsandstein, Muschelkalk) of the North German Basin. Hallesches Jb Geowiss B, Beih vol 5, pp 27–28
30 Schröder B (1982) Entwicklung des sedimentbeckens und stratigraphie der klassischen germanischen trias. Geologische Rundschau 71(3):783–794
31 Jublitz KB, Znosko J, Franke D (1985) Lithologic-palaeogeographic map. middle bunter, 1:1.500.000. International geologic correlation programme project no. 86, Southwest border of the East-European Platform, Zentrales Geologisches Institut, Berlin, G.D.R
32 Szulc J (2000) Middle triassic evolution of the northern peri-tethys area as influenced by early opening of the tethys ocean. Annales Societatis Geologorum Poloniae 70:1–48
33 Schwarz HU (1975) Sedimentary structures and facies analysis of shallow marine carbonates (lower muschelkalk, middle triassic, southwestern Germany). Contributions to Sedimentology, Stuttgart 3:1–100
34 Paul J, Franke W (1977) Sedimentologie einer transgression: die Röt/Muschelkalk-Grenze bei Göttingen. N Jb Geol Paläont Mh 3:148–177
35 Senkowiczowa H (1976) The Trias – The Polish lowlands. In: Geology of Poland 1: Stratigraphy, Part 2. Publishing House Wydawnictwa Geologiczne, Warsaw, pp 79–94
36 Brandner R (1984) Meeresspiegelschwankungen und tektonik in der trias der NW tethys. Jahrbuch für Geologie.A-B (Wien) 126:435–475
37 Maystrenko YP, Bayer U, Scheck-Wenderoth M (2005) The glueckstadt graben, a sedimentary record between the north and baltic sea in north central europe. Tectonophysics 397:113–126
38 Wurster P (1968) Paläogeographie der deutschen trias und die paläogeographische orientierung der lettenkohle in südwestdeutschland. Eclog geol Helv 61:157–166
39 Paul J, Wemmer K, Ahrendt H (2008) Provenance of siliciclastic sediments (Permian to Jurassic) in the central european basin. Zeitschrift der Deutschen Gesellschaft fuer Geowissenschaften 159(4):641–650
40 Ziegler PA (1988) Evolution of the arctic-north atlantic and the western thetys. aapg memoirs 43:198 p and 30 plates
41. Antrett P, Vackiner AA, Wollenberg U, Desbois G, Kukla P, Urai JL, Stollhofen H, Hilgers C (2011) Nano-scale porosity analysis of a Permian tight gas reservoir. In: Proceedings of extended abstract, AAPG international conference and exhibition. Milan, Italy
42. Sclater JG, Christie PAF (1980) Continental stretching: an explanation of the post-mid Cretaceous subsidence of the central North Sea basin. J Geophys Res 85:3711–3739
43. Schmoker JW, Halley RB (1982) Carbonate porosity versus depth; a predictable relation for South Florida. AAPG Bull 66:2561–2570
44. Watts AB (2001) Isostasy and flexure of the lithosphere. Cambridge University Press, Cambridge 458 p
45. Glennie KW (1986) Development of NW Europe‘s southern Permian gas basin. In: Brooks J, Goff JC, van Horn B (eds) Habitat of Paleozoic gas in N.W. Europe, vol 23. Geological Society of London, London, pp 3–22
46. Legler B. (2005) Faziesentwicklung im Südlichen Permbecken in Abhängigkeit von Tektonik, eustatischen Meeresspiegelschwankungen des Proto-Atlantiks und Klimavariabilität (Oberrotliegend, Nordwesteuropa): Schriftenreihe der Deutschen Gesellschaft für Geowissenschaften, vol 47. p. 103

47. Vackiner AA, Antrett P, Stollhofen H, Back S, Kukla PA, Bärle C (2011) Syndepositional tectonic controls and palaeo-topography of a Permian tight gas reservoir in NW Germany. J Pet Geol 34(4):411–428
48. Ziegler PA (1990) Geological Atlas of Western and Central Europe. Shell Int Pet Mij Geol Soc London 2:239
49. Scheck-Wenderoth M, Lamarche J (2005) Crustal memory and basin evolution in the Central European Basin System - new insights from a 3D structural model. Tectonophysics 397:143–165
50. Parnell J (2004) Titanium mobilization by hydrocarbon fluids related to sill intrusion in a sedimentary sequence, Scotland. Ore Geol Rev 24(1–2):155–167 (Ores and organic matter)
51. Jackson MPA, Vendeville BC, Schultz-Ela DD (1994) Structural dynamics of salt systems. Annu Rev Earth Planet Sci 22:93–117
52. Geluk M (1999) Late Permian (Zechstein) rifting in the Netherlands; models and implications for petroleum geology. Pet Geosci 5:189–199
53. Geluk MC (2000) Late Permian (Zechstein) carbonate-facies maps, the Netherlands. Geologie en Mijnbouw, Neth J Geosci 79(1):17–27
54. Schöner R, Gaupp R (2005) Contrasting red bed diagenesis; the southern and northern margin of the Central European Basin. Int J Earth Sci 94(5–6):897–916 (Dynamics of sedimentary basins; the example of the Central European Basin system)
55. Walsh P, Schultz-Ela DD (2003) Mechanics of graben evolution in Canyonlands National Park. Utah GSA Bull 115(3):259–270
56. Mohr M, Kukla PA, Urai JL, Bresser G (2005) Multiphase salt tectonic evolution in NW Germany; seismic interpretation and retro-deformation. Int J Earth Sci 94(5–6):917–940
57. Kukla PA, Urai JL, Mohr M (2008) Dynamics of salt structures. In: Littke R, Bayer U, Gajewski D, Nelskamp S (eds) Dynamics of complex intracontinental basins; the Central European Basin system. Springer, Berlin, pp 291–306
58. Mohr M, Warren JK, Kukla PA, Urai JL, Irmen A (2007) Subsurface seismic record of salt glaciers in an extensional intracontinental setting (late Triassic of northwestern Germany). Geology 35:963–966
59. Duval BC, Cramez C, Jackson MPA (1992) Raft tectonics in the Kwanza Basin, Angola. In: Jackson MPA (ed) Special issue salt tectonics, vol 9(4). Angola, Marine and Petroleum Geolog, pp 389–404
60. Ziegler PA (1995) Geodynamics of compressional intra-plate deformations: a comparison with the Alpine Foreland. Nova Acta Leopold, NF 71(291):265–300
61. Ziegler PA (1982) Geological Atlas of Western and Central Europe. Elsevier Science Ltd, Amsterdam, p 130
62. Lokhorst A (1998) The Northwest European Gasatlas. Netherlands Institute of Applied Geoscience TNO, Haarlem

Chapter 7
Facies Analysis from Well Cores, Northern Central Germany: Comparison to NW German Well Cores

7.1 Introduction

This chapter includes the sedimentary facies analysis of 4 wells (Well I–IV; Figs. 7.1 and 7.2), which are situated in northern central Germany, at the southern margin of the Southern Permian Basin (SPB).

In contrast to the north–western German cores from the main reservoir intervals of the Wustrow and Bahnsen Members (Upper Rotliegend II), this analysis focuses on younger stratigraphic intervals of the Bahnsen to Munster Members of the Upper Rotliegend II (Fig. 2.2). Another difference between the two study areas comprises the Upper Rotliegend II palaeo-geographical situation of the two study areas. In comparison to the north–western German study area, the northern central German study area was located closer to the perennial saline lake that occupied the central part of the SPB since the deposition of the Dethlingen Formation (Figs. 1.2, 7.1 and 2.2; [1]). Short-term, but widespread marine ingressions into the ephemeral shallow lake during the Niendorf and Munster Members of the Hannover Formation [2, 3] affect the sedimentation. Deposition is dominated by siliciclastics and subordinate evaporites [4–7] discussed ephemeral fluvial (wadi), aeolian, sabkha and lacustrine environments as the four major facies associations deposited at the SPB margins.

Results of the sedimentary facies analysis of both study areas were compared to improve the understanding of the depositional environment and to identify the sedimentary facies depending on the palaeo-geographic position at the southern margin of the SPB during the Upper Rotliegend II.

Applying the methodology, which was developed during the analysis of core material of the north–western German tight gas field (Sect. 3.1), the focus lay on macroscopic classification of sedimentary facies considering grain sizes, clay content, and sedimentary structures. Facies logs with grain size profiles were established and plotted in stratigraphic order: The lower left column shows the oldest stratigraphy, the upper right column the youngest (Figs. 7.3, 7.4, 7.5 and 7.7, 7.8). Each well is displayed as one profile with each core (Table 7.1) of a single well in a separate column and complementary core photos of characteristic

A. A. Vackiner, *Sedimentary Facies Reconstruction and Kinematic Restoration of Tight Gas Fields*, Springer Theses, DOI: 10.1007/978-3-642-36047-3_7,

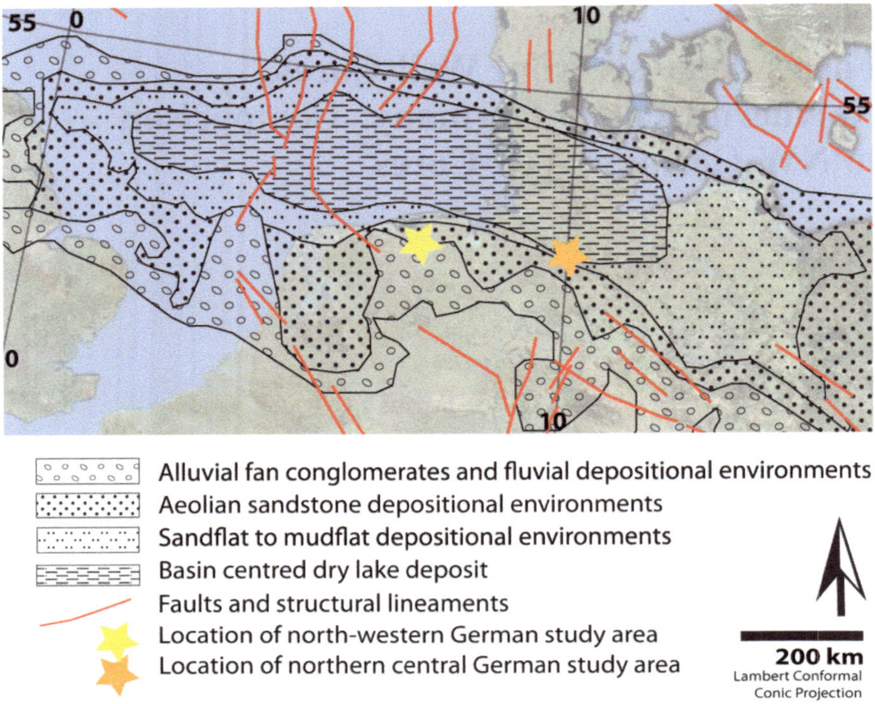

Fig. 7.1 Map outlining maximum extent of depositional areas of the SPB during the late upper Rotliegend II (modified from [7, 8]

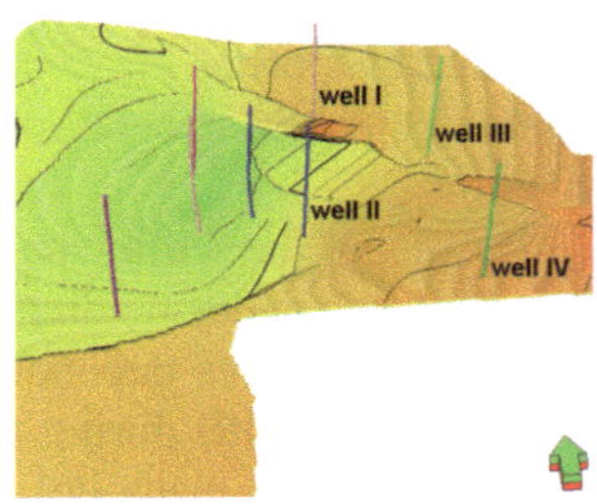

Fig. 7.2 Wells with analyzed core material depicted with top Rotliegend depth map

features. For the key to the profiles, including the colour code for the depositional environment and signatures for lithologies and syn- and post-sedimentary structures, see Fig. 7.3.

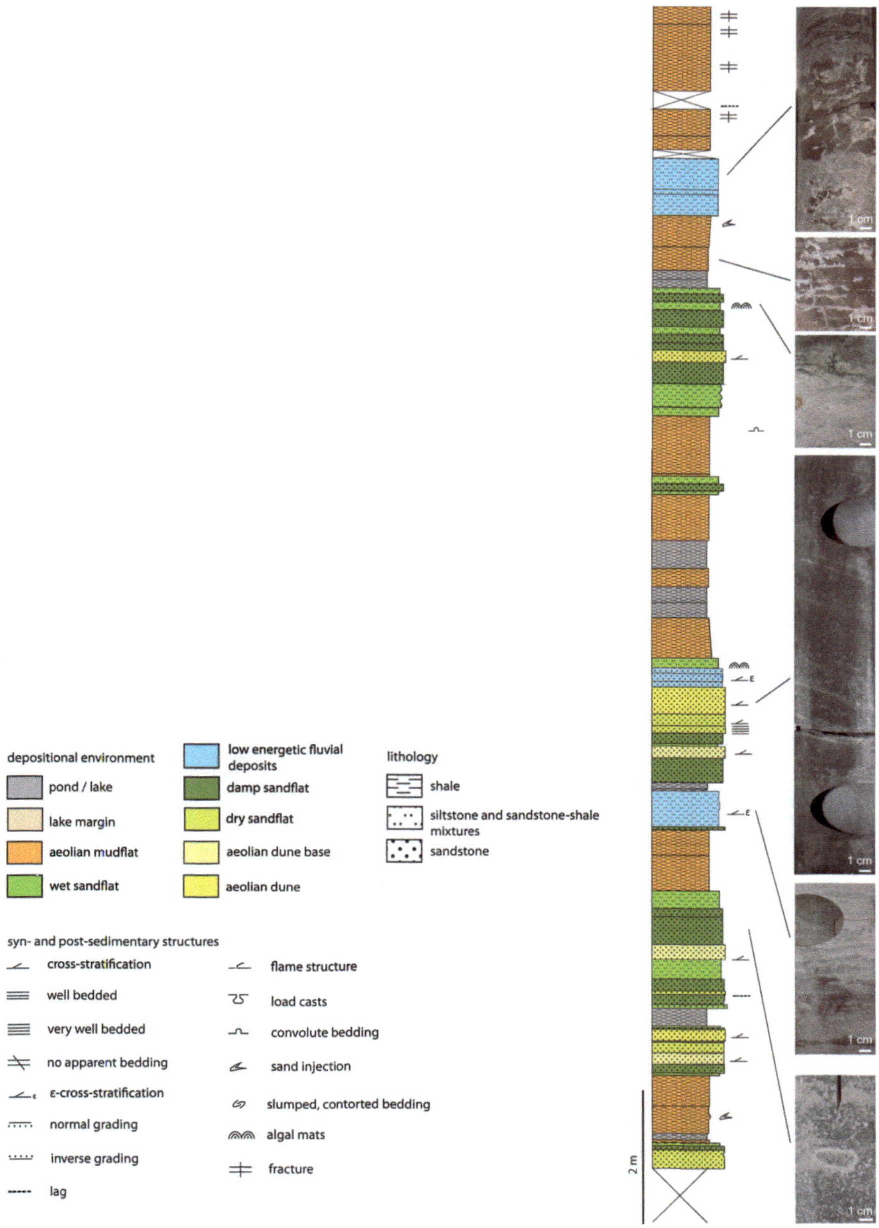

Fig. 7.3 Well I, core 4

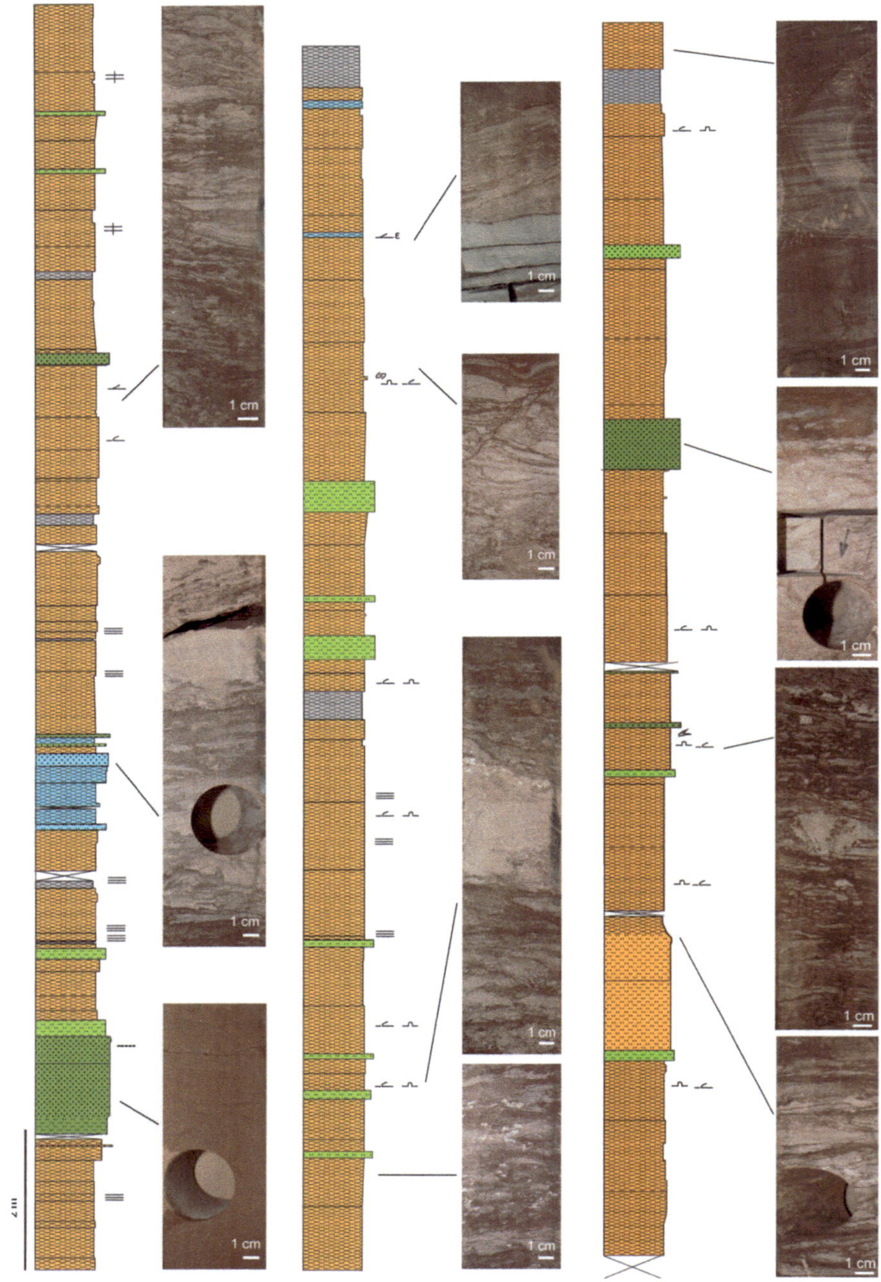

Fig. 7.4 Well I, core 3 to 1 (cf. p. 103 for key)

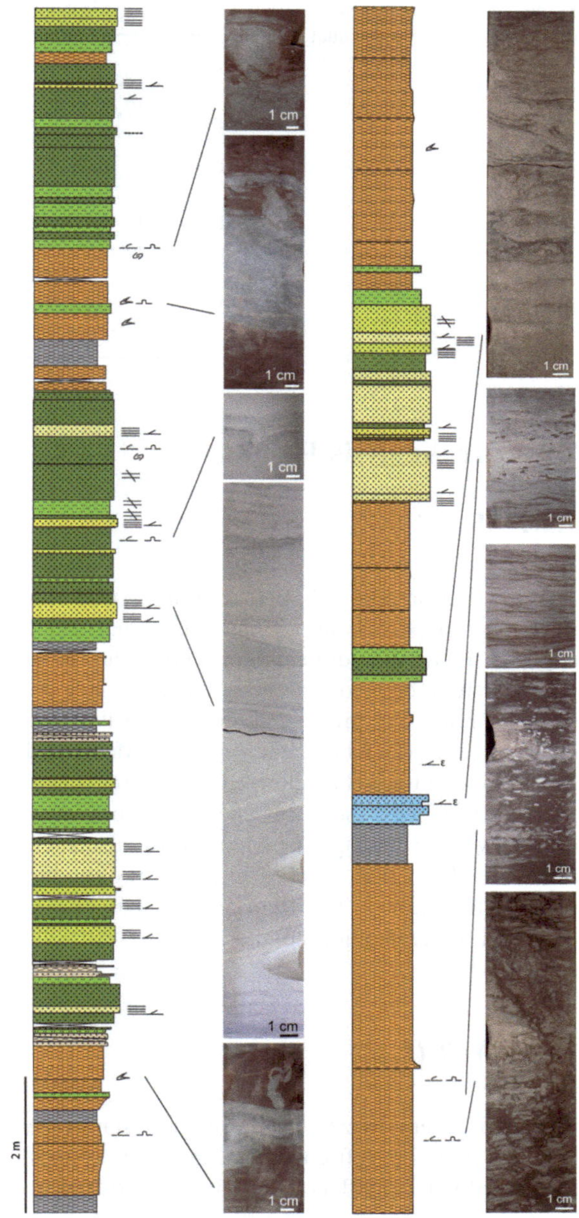

Fig. 7.5 Well II, core 2 and 1 (cf. p. 103 for key)

Table 7.1 Core-metres and stratigraphy of the individual Wells

	Core-metres	Formation/member
Well I	13.3	Hannover/Bahnsen—Munster
	18.2	
	17.6	
	18 m	
Well II	17 m	Hannover/Bahnsen
	18 m	Hannover/Niendorf—Munster
Well III	18 m	Hannover/Bahnsen—Munster
	18 m	
	18 m	
Well IV	12.6 m	Hannover/Niendorf
	12.6 m	Hannover/Bahnsen

7.2 Core Analysis Results

7.2.1 Well I

The lowermost core of Well I (Bahnsen Member; Fig. 7.3) is dominated by sandstones deposited as aeolian dunes and sandflats. Prevailing damp sandflat deposits indicate sedimentation under the influence of a relatively shallow groundwater table, close to the sediment surface. Pond and mudflat deposits in this succession are interpreted interdune deposits. To the top, predominantly mudflat deposits occur (Dambeck and Niendorf Member; Fig. 7.4). They developed sub-aquatically and are characterized by varying silt to fine grained sandstone lenses. In some instances, the occurrence of ε-cross stratification and rip-up clasts indicates fluvial reworking. The Munster Member includes a damp aeolian dominated succession, which is encased in prevailing mudflat deposits of under- and overburden stratigraphy. The Munster Member sandstones are bleached and contain abundant mineralization nodules.

7.2.2 Well II

Well II is characterized by the highest sandstone content of the analyzed core material (Fig. 7.5, left column). Most of the sandstone and siltstone intervals are associated with sandflats, which were deposited in a temporary wet environment, respectively under the influence of a shallow groundwater table. Several sandflats are classified as fluvially reworked, due to the occurrence of clay rip-up clasts. Aeolian dunes or dry aeolian sediments are subordinate facies. The occurrence of mineralization nodules and early diagenetic concretions indicates a diagenetic overprint of the sediments (Fig. 7.6). The uppermost core section is mainly dominated by mudflat and subordinate wet sandflat deposits (Fig. 7.5, right

Fig. 7.6 Diagenetically overprinted dune deposit, picture from Well II

column). Some mudflat to wet sandflat successions include rip-up clasts and are therefore interpreted as being fluvially reworked. The deposition of the mudflats took place sub-aquatically with only short-term drying events, indicated by the intercalation of damp sandflats, during which the sedimentation was influenced by a shallow groundwater table.

7.2.3 Well III

In the lower part of Well III damp sandflat deposits are the prevailing sedimentary facies (Fig. 7.7, left column). The groundwater table was located close to the sediment surface during sedimentation. The occurrence of different types of potential diagenetic nodules and bleached horizons is a characteristic for the macroscopic identification of diagenetic overprint. To the top, the core is

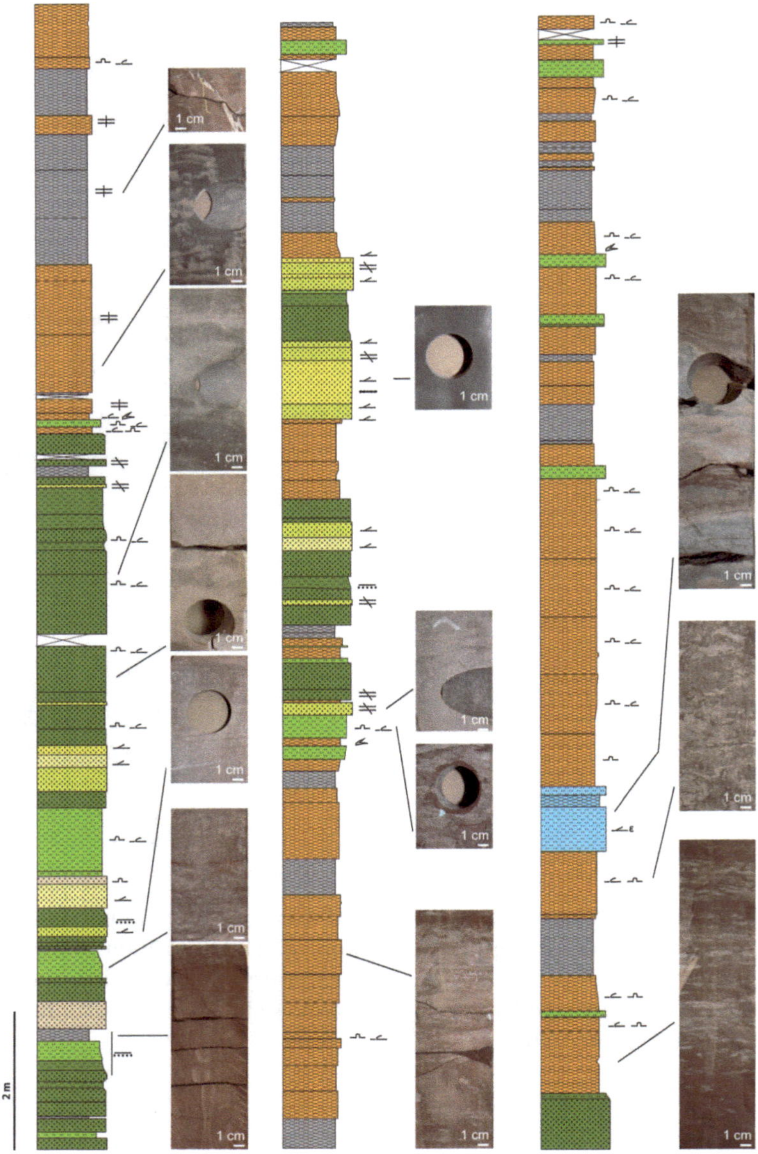

Fig. 7.7 Well III, core 3 to 1 (cf. p. 103 for key)

dominated by sub-aquatically deposited sediments, which encase two aeolian successions (Fig. 7.7, upper left column). Dip directions of the lower dune sets vary and can therefore be classified as barchanoid dune system. The upper aeolian succession mainly consists of sandflat deposits with shallow dip angles. Abundant damp sandflat deposits indicate that the groundwater table was located close to the sediment surface during deposition. Interdune deposits are characteristic for temporarily wet conditions. A wetting upward trend can be observed for the overlying sedimentary facies. Minor wet sandflat deposits with high clay content are encased in clay dominated successions. The occurrence of rip-up clasts and ε-cross-stratification indicates fluvial reworking. Abundant post-sedimentary structures e.g., convolute bedding and flame structures in the mudflat and wet sandflat deposits are indicative for dewatering and desalinization (Fig. 7.7, right columns).

7.2.4 Well IV

The lower part of Well IV predominantly consists of mudflat and pond to lake deposits (Fig. 7.8, right column). The clay-dominated sediments show abundant horizontal and minor vertical fractures. Minor coarser grained areas are fluvially influenced. The upper part of the cored section contains several sandflat to aeolian dune deposits (Fig. 7.8, left column). The Munster Member, in the upper part of the cored section, contains damp sandflat to aeolian dune deposits of higher primary maturity, but shows a significant diagenetic overprint.

7.3 Interpretation

Significant aeolian successions mainly occur in the Bahnsen Member and Munster Member. These aeolian intervals were deposited in a temporarily wet environment, indicated by the occurrence of several wet interdune areas. Even during the deposition of aeolian dunes, the groundwater level is assumed to have maintained close to the sediment surface. The aeolian sediments of the Munster Member have undergone a higher diagenetic overprint, indicated by whitish mineralization nodules (anhydrite, minor calcite) and diagenetic concretions (e.g., in the lower part of Well II; Fig. 7.3). Furthermore, the sediments of the Munster Member are partly bleached and abundant fractures, which are cemented by mica and quartz, offset the primary sedimentation structures (cross bedding, layered deposits).

In general, the largest aeolian interval is located in the Bahnsen Member of the Well II. Because of the limited thickness of aeolian successions, a distinct differentiation of barchanoid dune and amalgamated dune deposits throughout a well is not possible. In Well III a barchanoid dune set was identified. In general, the dunes are interpreted to have been limited in height, not exceeding 20 m, and to have been accompanied by multiple interdune deposits. Dunes were rather of barchanoid dune type, including extensive interdune areas. Stacked dune sets are not present in the study area.

Fluvial deposits were generally sedimented under low energy currents. They, most likely, originate from low energy sheetflood deposits discharging into lake or

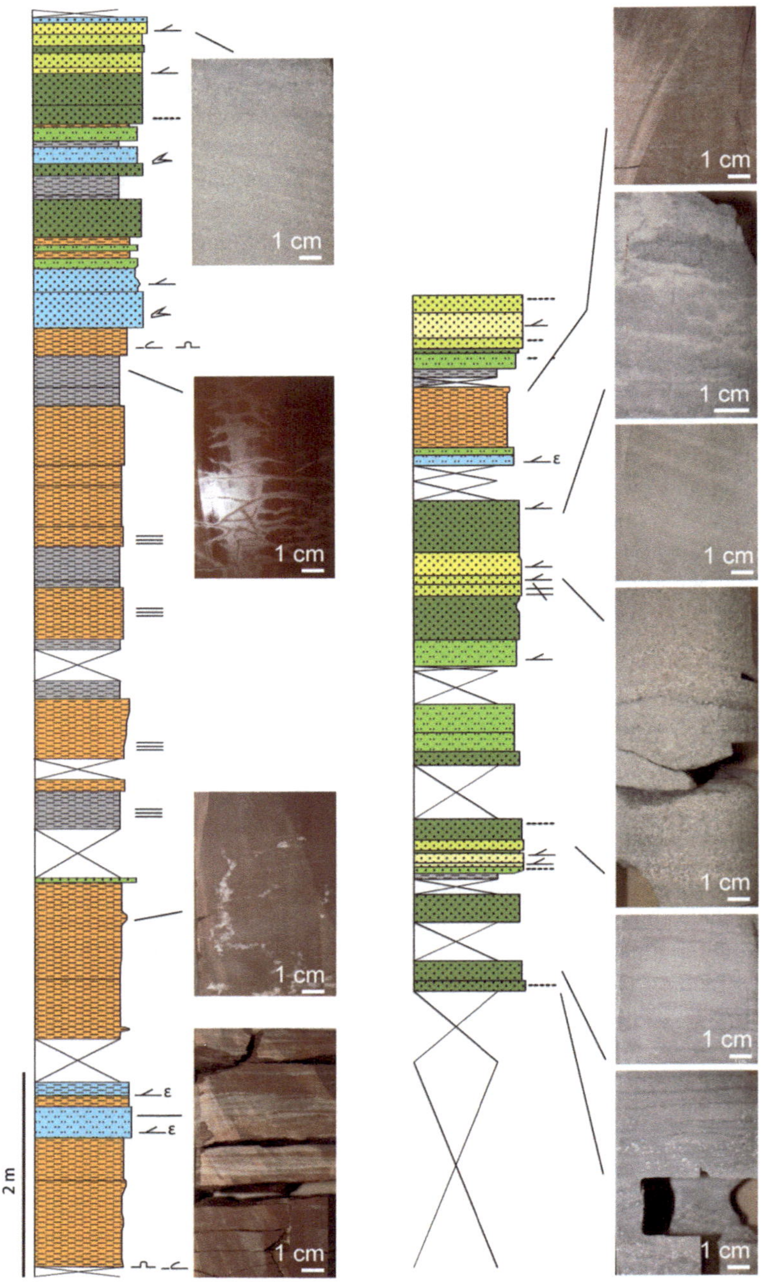

Fig. 7.8 Well IV, core 2 to 1 (cf. p. 103 for key)

pond environments. Fining upward (FU) cycles with medium grained sandstone bases were interpreted in the Wells I and III. These indicate the existence of very small meandering fluvial channels with depth ~0.5 m and widths ~0.5–3 m. Braided stream deposits were not identified. The lack of erosional structures and limited grain sizes points to low palaeo-topographic gradients. However, grain sizes are also dependant on the source area and reworking processes.

The sedimentation was mainly governed by sub-aquatic deposition. Associated sedimentary facies are mudflats, pond and lake deposits. Damp and wet sandflat deposits developed during times of a shallow groundwater table and ephemeral flooding of the sediment surface. The clay intervals contain micro-fracture systems which are mainly oriented horizontally, most likely layer bound, but are in some cases also vertically connected. Salt efflorescence along the fracture network is common. The observation of such fractures restricted to the clay intervals might be essential for future hydrocarbon field development including the application of hydraulic fracturing techniques.

Taking the palaeo-geographic situation of the wells and the extension of the Southern Permian Basin into account, the following conclusions can be drawn:

1. Well IV is characterized by minor influence of fluvial sedimentation, most likely with sediment source in the south.
2. All wells are located in an area of aeolian sedimentation during the Bahnsen and partly during the Munster Member. Due to the palaeo-geographic situation just south of the flooded great saline lake area of the SPB, the sedimentation was influenced by a shallow groundwater table. The dominant sediments in these aeolian successions are therefore damp sandflat deposits. Episodic low energetic fluvial currents and lake highstands influenced the sedimentation.
3. Because of the prevailing existence of mudflats during the intermediate Members of the Hannover Formation (Dambeck and Niendorf Formation), the study area was most likely flooded and only ephemerally dry during the deposition of the analyzed intervals.

7.4 Comparison to North–Western German Study Area

The northern central German area is located ~200 km east of the north–western German tight gas field, in an area that was situated closer to the centre of the SPB during the Upper Rotliegend II (Figs. 1.2 and 7.1). In contrast to the north–western German study site, the analysis of core material concentrates on the Bahnsen Member to Munster Member of the Hannover Formation and not on the main reservoir interval of the Wustrow and Bahnsen Member. As a result, the area was affected by lake level changes and marine ingression into the saline lake that occupied the centre of the SPB since the Dethlingen Formation.

Fluvial deposits in the northern central German study area are interpreted as low energetic flashflood deposits with very little to no depositional slope gradient.

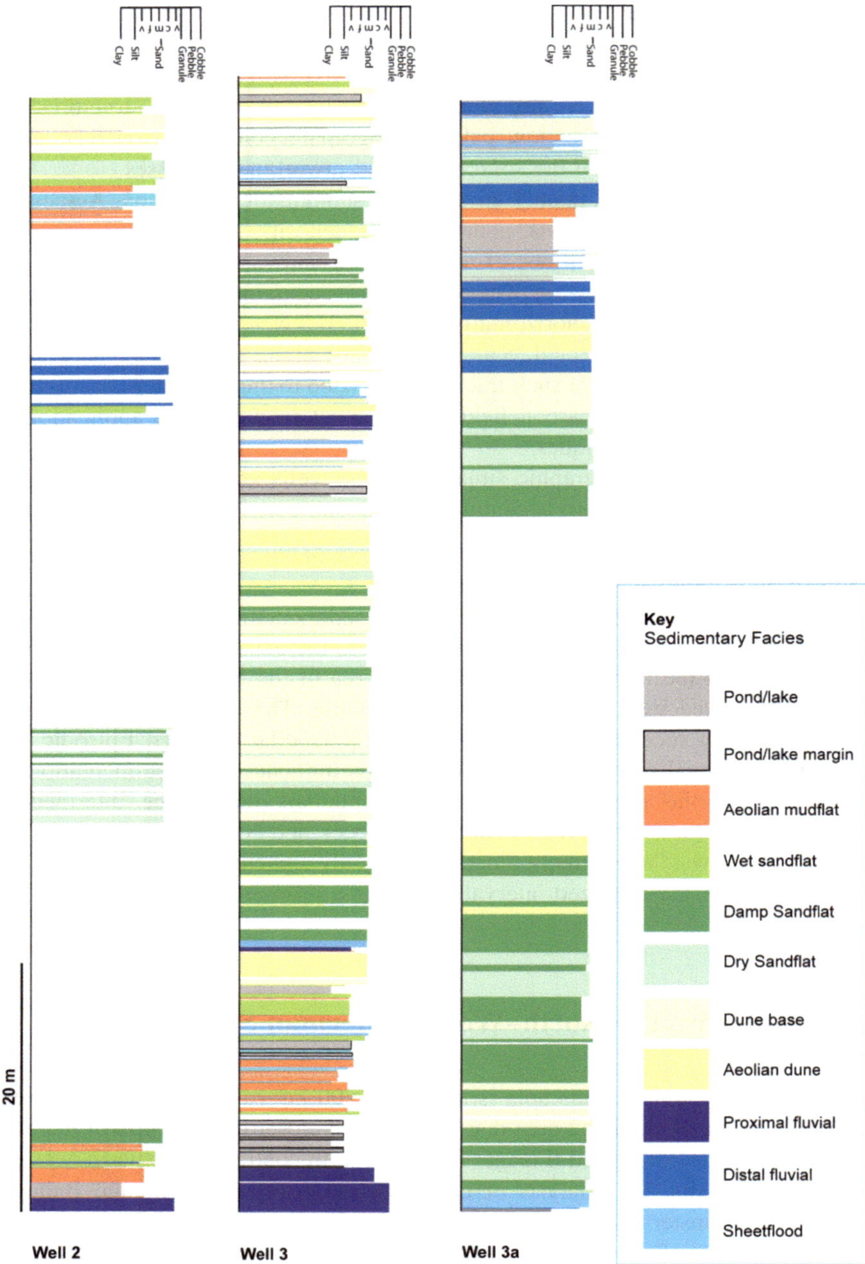

Fig. 7.9 Facies analysis profiles of the North–Western German Well cores

In contrast, in the north–western German study area, conglomeratic braided stream depositional systems at the base of the Wustrow Member indicate that a certain palaeo-topography gradient was present during deposition (Fig. 7.9; cf. Chap. 4).

The core material of the northern central German study area includes only minor successions of aeolian deposits, mainly concentrated on the Bahnsen and Munster Member, which developed during regressions of the perennial saline lake. Isolated barchanoid dunes, encased in clay rich deposits, were identified. The sedimentary facies associations mainly indicate sub-aquatic deposition with episodic short-term subaerial exposure. A high saline content in the clay dominated intervals of the cores is characteristic for periodical wetting and drying cycles (Fig. 2.2). In contrast, the core material of the north–western German study site contains temporarily wet deposits as well (up to 50 % of core material), but dry to damp aeolian deposition during the Wustrow and Bahnsen Members was prevailing (Fig. 7.9). Abundant aeolian dune deposits allow a classification of dune systems into barchanoid and aklé dunes. Dune systems are not only laterally attached, they are occasionally vertically stacked. In terms of reservoir conditions, the north–western German study area with the dry aeolian successions provides the better primarily gas reservoir.

References

1. Gast R, Gaupp R (1991) The sedimentary record of the late Permian saline lake in N.W. Germany. In: Renaut RW, Last WM (eds) Sedimentary and paleolimnological records of Saline lakes. Natl Hydrol Res Inst, Saskatoon, Canada, pp 75–86
2. Gast R, Gebhardt U (1995) Elbe Subgruppe. In: Plein E (ed) Stratigraphie von Deutschland I; Norddeutsches Rotliegendbecken: Rotliegend-Monographie Teil II, vol 183. Courier Forschungsinstitut Senckenberg, pp 121–145
3. Legler B, Gebhardt U, Schneider JW (2005) Late Permian non-marine: marine transitional profiles in the Central Southern Permian Basin. Int J Earth Sci 94:851–862
4. Glennie KW (1972) Permian Rotliegendes of Northwest Europe interpreted in light of modern desert sedimentation studies. AAPG Bulletin 56:1048–1071
5. George GT, Berry JK (1993) A new palaeogeographic and depositional model for the upper Rotliegend of the UK sector of the Southern North Sea. In: North CP, Prosser DJ (eds) Characterization of Fluvial and Aeolian Reservoirs, vol 73. Geological Society of London, Special Publication, pp 291–319
6. Strömbäck AC, Howell JA (2002) Predicting distribution of remobilized aeolian facies using sub-surface data: the Weissliegend of the UK Southern North Sea. Pet Geosci 8:237–249
7. Legler B (2005) Faziesentwicklung im Südlichen Permbecken in Abhängigkeit von Tektonik, eustatischen Meeresspiegelschwankungen des Proto-Atlantiks und Klimavariabilität (Oberrotliegend, Nordwesteuropa): Schriftenreihe der Deutschen Gesellschaft für Geowissenschaften 47:103
8. Ziegler PA (1982) Geological atlas of Western and Central Europe: Elsevier Science Ltd., p 130

Chapter 8
Conclusions, Synopsis and Perspectives

8.1 Conclusions

The questions that built the framework for this study, described in the Introduction, are discussed in the following.

How was the Upper Rotliegend II fault-induced palaeo-topography arranged prior to later multi-phase tectonic overprint?

The study area records evidence for tectonic activity contemporaneous with Upper Rotliegend II deposition visible on isopach maps. The reconstructed graben structure is characterized by bounding N–S-trending fault zones with offsets of up to 250 m in the W (along FZ-4) and up to 150 m in the E (along FZ-1). To the N, the eastern fault zone (FZ-1) terminates, and the asymmetric graben changes into a half-graben. The maximum palaeo-relief, induced by fault activity, during the deposition of the Upper Rotliegend II, is estimated 250 m along the western fault zone (FZ-4). Along its hangingwall an Upper Rotliegend II depocentre with sedimentary thickness maxima of 450 m can be identified. The reconstructed graben to half-graben itself is partly subdivided by a central fault zone (FZ-3), which consists of three westward dipping faults during the Upper Rotliegend II. Fault-controlled palaeo-relief with $\sim$100–150 m height has been estimated only for the westward-dipping central fault of the fault zone. Within the N–S-oriented fault zones dip-slip and oblique-slip in a left-lateral transtensional regime have been considered. Local depocentres that are associated with complex relay ramps and pull-apart sub-basins induced by left-lateral transtensional stress provide accommodation for later hydrocarbon reservoir rocks.

N–S oriented dip-slip faults with a certain oblique-slip component are generally characterized by synsedimentary activity during the entire Upper Rotliegend II. NNW-SSE-trending faults, in most instances, originate from later connection of N–S-trending fault segments. Consequently, palaeo-relief that developed during the Upper Rotliegend II can be regarded as fault-induced. During the Upper Rotliegend II NW–SE-oriented faults did not have synsedimentary fault-induced palaeo-relief developed. They formed under later stress regimes.

A. A. Vackiner, *Sedimentary Facies Reconstruction and Kinematic Restoration of Tight Gas Fields*, Springer Theses, DOI: 10.1007/978-3-642-36047-3_8,

The Altmark IV tectonic event, which comprised normal faulting during deposition of the Upper Rotliegend II Ebstorf and Wustrow Members, is identified as trigger for the onset of deposition in the study area. Later, during the Upper Rotliegend II, more localized tectonic phases caused additional fault offsets. Latest Rotliegend fault activity on N–S-trending faults is evident in the north-western German area of East Frisia.

Where are the synsedimentary Upper Rotliegend II structural traps for the accumulation of reservoir rocks located in the study area?

Fault-induced topography serves as crucial parameter in the analysis of sedimentary facies distribution. Topographic highs are exposed to high wind velocities and serve as barriers to wind transported material, which deposits along the lee and windward side. Dune accumulation and preservation potential is thus favoured in hangingwall positions, whereas palaeo-highs may provide early hydrocarbon traps. Larger thicknesses of dune sand accumulation and therefore of gas reservoir rocks are anticipated in palaeo-hangingwall positions. Furthermore, local depocentres, which provide accommodation for clastic reservoir rocks comprising sandstones and conglomerates during the Upper Rotliegend II are associated with localized pull-apart (FZ-1, FZ-4) and overstep structures, at releasing locations of strike-slip and oblique-slip faults (FZ-3), suggesting left-lateral transtensional stresses. Depending on fluctuations of the groundwater table the relative basin areas might have undergone ephemeral flooding events. As fault-induced palaeo-relief is mainly postulated for the N–S-trending faults and fault segments of fault zones, high thickness sandstone reservoir rock deposits are anticipated to be mainly located in hangingwall positions of these faults. According to the model, NW–SE-oriented fault zones without fault-induced topography during the Upper Rotliegend II, would encounter less thick sandflat deposits. The exceptional and localized preservation of dune deposits in palaeo-footwall positions (FZ-1) may relate to subtle palaeo-relief, the formation of sub-seismic scale footwall collapse compartments and increased moisture content at the sediment surface.

To further extend the accommodation analysis, a field analogue study with focus on fault activity, controlling the sedimentary facies distribution and sediment-fault interactions, was carried out. Sandflat deposits and aeolian dune deposits are regarded as the reservoir rocks of highest quality in the tight gas study area. Based on the field analogue study in the Panamint Valley, CA, USA, a model of topography, synsedimentary fault activity, sediment source areas, and dominant wind direction as key controlling parameters of the sediment facies distribution has been developed and compared to the north-western German subsurface study site reconstructed to the Upper Rotliegend II setting prior to multi-phase tectonic overprinting. In both study sites, dunes are situated on braided stream alluvial fans in mid slope position. In the north-western German study site, dunes on the footwall of the eastern fault zone (FZ-1) developed as shallow barchanoid to aklé dunes from a unimodal E/ENE wind direction in the upper position of a synsedimentary leeside trap. The abrasion of quartz-grain coatings on the footwall of the fault zone is a proxy for an active dune system. Thicker alluvial and aeolian deposits are associated with the lower part of the leeside trap on the hangingwall of the fault zone (FZ-1). The western fault zone (FZ-4) represents a windward trap for aeolian sediments with sandflat deposits in the exposed, high wind velocity zone on the footwall. Higher amounts of alluvial and aeolian accumulation are considered to be present on the hangingwall.

Can the Upper Rotliegend II sedimentary facies distribution be reconstructed using a field analogue study?

The comparison of the two study sites shows that the Panamint Valley represents a well-suited modern analogue for the German subsurface tight gas reservoir during the Upper Rotliegend II. Especially the position of the aeolian sandstone deposits, comprising dune and sandflat deposition controlled by fault-induced topography that acts as windward and lee side trap for aeolian sand, is of high similarity in both study sites. The field analogue observations concerning abrasion of quartz grain coatings, dune types and sizes, the presence or absence of desert varnish, and the incision depth of alluvial fan channels have been transposed to reconstruct the sediment dynamics of the north-western German study area during the Upper Rotliegend II. The deepest Upper Rotliegend II basin area of the asymmetric graben to half-graben is located on the hangingwall of the western fault zone (FZ-4) of the German subsurface study site. Compared to the Panamint Valley dry lake location in the deepest basin area, I postulate that it represented a possible location for the development of dry to ephemeral lakes, or was at least exposed to the influence of the groundwater table. This configuration potentially favoured the deposition of high aeolian dunes, stabilized by damp dune bases, in a windward trap.

In general, the study shows that a well-suited field analogue study enables (1) a detailed interpretation, interpolation, extrapolation, and prediction of a sedimentary facies distribution for only spatially limited core data across larger areas, (2) a transfer of key mechanisms of sedimentary facies distribution to subsurface data using fault interpretation, retro-deformation and palaeo-relief reconstruction, and (3) the reconstruction of sediment dynamics in tectonically overprinted subsurface areas by detailed observations in the field and the determination of well-suited sediment characteristics.

What role does halokinesis and multi-phase tectonic overprinting play during reservoir rock development?

Upper Rotliegend II and/or older faults were repeatedly reactivated under changing stress regimes, in particular during Triassic-Jurassic faulting, often causing the development of cumulative fault offsets and a post-depositional fault propagation. In most instances, the mainly N–S-trending Rotliegend faults (e.g., FZ-1 and its suspected northern continuation by FZ-5 and FZ-3) successively enlarged their extents and finally connected. The present-day structural configuration records the cumulative effects of successive multi-tectonic overprint of deformation phases during the Rotliegend, Zechstein, mainly Triassic, Jurassic and Cretaceous. To quantify these phases of multi-tectonic overprinting, a sequential retro-deformation with main focus on salt tectonics, was carried out.

Analogue modelling studies suggest that salt cannot transmit large differential stresses due to its low viscosity [1]. Therefore, post-salt series are tectonically decoupled from the substratum by the intermediate Zechstein salt acting as a décollement surface. The pre-salt reservoir rock interval of the Upper Rotliegend II shows diagenetic overprinting, which degraded the reservoir sandstones, and can be correlated with the major tectonic phases. The Zechstein salt underwent different salt rise mechanisms, which are highly affected by the different tectonic phases.

Three phases of salt deformation mechanisms can be subdivided and correlated to distinct diagenetic overprinting phases: (1) Salt movements started with lateral salt flow. Salt diapirism and rafting were initiated by tectonically induced differential loading during thin-skinned extension. Extensional tectonics caused small salt injections into the overburden. Some injections grew to diapirs during a raft tectonic phase in the Upper Buntsandstein and Muschelkalk. Mineral reactions during this phase did not degrade the compositionally and texturally mature gas reservoir sandstones of the tight gas field. (2) Keuper to Jurassic were dominated by active salt diapirism, triggered by regional extension with a maximum extension rate during Jurassic. The Jurassic peak temperatures, which accompanied the change from active to passive diapirism, triggered quartz overgrowth and the formation of authigenic clay minerals [2]. The salt pressure exceeded the brittle strength of the overburden during active diapirism. The geometry of the diapiring salt wall with its Keuper and Jurassic salt namakiers is shaped by several periods of subaerial exposure of salt, occurring when the rate of upwelling was higher than the rate of dissolution, aggradation and extension. (3) Since the Lower Cretaceous, the diapiring salt wall was rising due to passive diapirism and extensive salt rim syncline development. This phase was influenced by further quartz overgrowth and clay mineral precipitation and underwent late-stage pore filling of sulphates and carbonates [2].

Are the developed methods applicable to geologically similar study areas?

The analysis of the well core material of an additional study area in northern central Germany, which is located 200 km E of the main study area, was carried out using the facies analysis methods developed during the study on the main north-western German study area (cf. Sect. 3.1). Even though the northern central German study area was located closer to the centre of the Southern Permian Basin (SPB), which was occupied by a saline lake during the Upper Rotliegend II, and the study concentrates on younger stratigraphy, similar fluvio-aeolian to ephemeral lake settings are observed. However, the amounts of the different facies associations considered are diverging. The dominant sedimentary facies in the northern central German study area are aeolian mudflat to wet and damp sandflat deposits. In contrast, the average moisture content in the north-western German study area was lower and mainly aeolian wet to dry sandflats and dunes were deposited.

In addition to the facies analysis of core material, one seismic line was depth converted on the basis of well records. Seismic interpretation revealed a huge salt namakier overlying Base Cretaceous. The sequential retro-deformation of this seismic line of the northern central German study area indicates the same salt deformation mechanisms as interpreted for the north-western German study area. Nevertheless, several significant differences occur comparing the timing of salt kinematics of the two study areas. In the north-western German study area, salt namakiers are mapped at the Base to Top Upper Keuper, whereas salt namakiers in the northern central German study area are interpreted at the Base Cretaceous. The subaerial exposure of salt was thus later in western Germany. Another difference comprises the development of salt rim synclines. Salt rim synclines in the north-western study area concentrate on the (Lower) Cretaceous indicating the beginning of passive diapirism/downbuilding. In the northern central German study area rim synclines already started to develop during the Jurassic. The Jurassic period, which is characterized by subaerial exposure of salt, likely marks the beginning of the down building phase in northern central Germany.

8.2 Synopsis

This study improves the overall understanding and localization of the Upper Rotliegend II tight gas reservoir rock facies.

Unprecedented insights into the detailed Upper Rotliegend II palaeo-topography and local tectonically induced sediment thickness changes prior to a multi-phase tectonic overprinting are one of the key results. The study further concentrates on the tectonically induced synsedimentary facies distribution in transtensional continental settings on the basis of a comparison with a modern field analogue, which enables a detailed analysis of the reservoir rock's distribution and its properties. Accommodation for the deposition of fluvio-aeolian reservoir rock facies, comprising alluvial fan, dune and sandflat deposits, are evaluated. The study is completed with the analysis of the influence of the multi-phase tectonic overprinting on the primarily (during Upper Rotliegend II) mature reservoir rocks. The regional tectonic events were therefore correlated with local tectonically induced salt kinematics and with diagenetic phases.

The outcome of this thesis shows that the unravelling of the complex reservoir architecture of deeply buried tight gas fields in central Europe, with its laterally adjacent sedimentary facies changes that underwent multi-tectonic and -diagenetic overprinting, needs multi-disciplinary research for de-risking tight gas exploration and production.

8.3 Perspectives

In this study the structural analysis mainly concentrated on the fault-induced sedimentary facies distribution. Synsedimentary faults also comprise zones of enhanced hydraulic activity [3, 4], which favour diagenesis, mineralization and fluid flow. Predictive identification of fluid pathways is of high interest for the

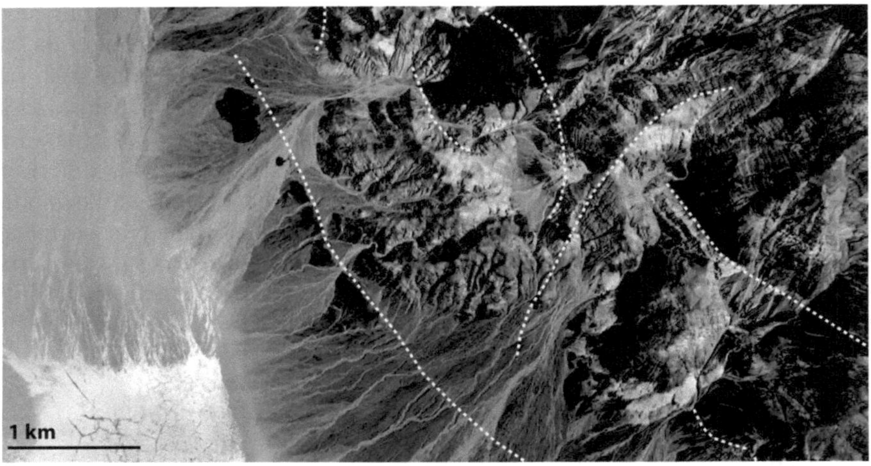

Fig. 8.1 Euhedral calcite precipitations visible as light areas on Satellite Image (copyright by Google and Digital Globe) of north-eastern Panamint Valley, faults pointed out by *white dotted lines*

predicting the distribution of tight gas reservoir rock quality. Enhanced fluid flow favours cementation, and may lead to field compartmentalization (e.g., [5]).

The field analogue study in the Panamint Valley provides important information about possible subsurface pathways that are open to fluid circulation. Fault striations along cementation of visible fault scarps (Fig. 5.3) and sulphur smells suggest recent fluid circulation in open faults. Precipitation of euhedral calcite crystals of up to 0.5 cm diameter can be observed along fault scarps, but they are also spread along certain stratigraphic layers, like limestone and mudstone or dolomite intervals, in the closer vicinity of faults (Fig. 8.1). A future study should concentrate on whether the tight gas reservoir rock, which was deposited in a Panamint Valley-similar environmental and geological setting, also underwent early degradation by synsedimentary fluid flow.

In addition a future study should focus on the differences in the timing of salt kinematics in different areas of the SBP.

References

1. Vendeville BC, Ge H, Jackson MPA (1995) Scale models of salt tectonics during basement-involved extension. Pet Geosci 1:179–183
2. Havenith VMJ, Meyer FM, Sindern S (2010) Diagenetic evolution of a tight gas field in NW Germany. DGMK/ÖGEW-Frühjahrstagung 2010, Fachbereich Aufsuchung und Gewinnung, Celle
3. Gaupp R, Matter A, Platt J, Ramsayer K, Walzebuck JP (1993) Diagenesis and fluid evolution in deeply buried Permian (Rotliegende) gas reservoirs NW Germany. AAPG Bull 77(7):1111–1128

4. Clauer N, Zwingmann H, Chaudhuri S (1996) Isotopic (K-Ar and oxygen) constraints on the extent and importance of the Liassic hydrothermal activity in Western Europe. Clay Miner 31:301–318
5. de Medeiros WE, do Nascimento AD, Antunes AF, de Sá EFJ, Neto FFL (2007) Spatial pressure compartmentalization in faulted reservoirs as a consequence of fault connectivity: a fluid flow modelling perspective, Xaréu oil field, NE Brazil. Pet Geosci 13:341–352

MIX
Papier aus verantwortungsvollen Quellen
Paper from responsible sources
FSC® C105338

If you have any concerns about our products,
you can contact us on
ProductSafety@springernature.com

In case Publisher is established outside the EU,
the EU authorized representative is:
Springer Nature Customer Service Center GmbH
Europaplatz 3, 69115 Heidelberg, Germany

Printed by Libri Plureos GmbH
in Hamburg, Germany